I0422399

Forging A More Perfect Union

BOOKS ON DIALOGUES FOR THE HUMAN POSSIBILITY

By
Kenneth R. Schneider

1. **Autokind Vs. Mankind:** An Analysis of Tyranny, A Proposal for Rebellion, A Plan for Reconstruction

2. **Destiny of Change:** How Relevant is Man in the Age of Development?

3. **On the Nature of Cities:** Toward Enduring and Creative Human Environments

4. **Forging a More Perfect Union:** For a Grand Harmony of Cities, Democracy, Ecology

5. **The Runaway Economy:** The Rhetoric is Growth, The Issue Is Freedom

6. **American Communities:** The Next Human Advance, A New Class of Freedom

7. **Shaping a Supremacy of Person:** For a Fundamental, Democratic Liberation

Forging A More Perfect Union

For a Grand Harmony of Cities, Democracy, Ecology

Kenneth R. Schneider

iUniverse, Inc.
New York Lincoln Shanghai

Forging A More Perfect Union
For a Grand Harmony of Cities, Democracy, Ecology

All Rights Reserved © 2004 by Kenneth R. Schneider

No part of this book may be reproduced or transmitted in any form or by any means, graphic, electronic, or mechanical, including photocopying, recording, taping, or by any information storage retrieval system, without the written permission of the publisher.

iUniverse, Inc.

For information address:
iUniverse, Inc.
2021 Pine Lake Road, Suite 100
Lincoln, NE 68512
www.iuniverse.com

ISBN: 0-595-33817-8

Printed in the United States of America

CONTENTS

PREFACE

In our history of material progress, development has centered upon science, technology, and business initiatives. However, society has advanced to such a point that material plenty can now be assumed, permitting a new outlook to take hold of our imagination. That new vision, I believe, must be social, that is, totally human in purpose and program. And, if it is to be social, it must function close to the individual person, for the only rightful focus of "a more perfect union" is to have vital human meaning in the twenty-first century. Ultimately there can be no higher purpose than the person.

It is in the spirit of the person that we can generate many thousands of inspirations, and perhaps only there can society rise to the range of human creativity that first began to take hold upon the ancient Greeks two and a half millennia back. And only there will we strike the chord that set upon "we the people" over two centuries ago.

If we focus on the person, I doubt that we can do better than set our minds and inspirations on the age-old community, give it modern organized strength, and direct our visions of good living onto its interpersonal level of life in which the people can place their trust, enthusiasm, and common commitment. It is there that association can lead to common enterprises of common interest and encourage people to inspire mutual growth, perhaps in the end to a new vision of life in society. If people are to prosper as people, an institution like community may become their chief instrument. And if communities can become vital for a significant population, they can also become the instrument to help renew the American city, since traditionally the city has been both generator and repository of civilization.

But today when the flood of materialism routinely outstrips our imagination of the good life, I write about decadent cities. I write about their dark side, not merely to press upon possibly the greatest defeat of modern development, but to lay the groundwork in later chapters for a vision to build a new and vast human potential of socially creative, democratically expansive, and ecologically valid cities—the highly achievable grand harmony. My starting point is that the engorgement of goods—and the moral strength of goods today being the primary goal of society—is creating corruptive and deadly cities. In our tragic historic blindness, there is but little recognition that cities are central to all that happens in economics, ecology, democracy, and the many rich, unrealized expanses of social life. Very simply, cities, perhaps society's most operative mechanism and the individual's critical environment for living both vigorously and graciously, are being defeated at a time of their greatest potential.

It is a cliché to say that we live in a vast technological revolution. Yet we do not recognize in any intimate way how profoundly that revolution now dominates thought and behavior, from the highest national policies to the purchase of a new house. And clearly, the revolution is still far from spent. Yet I submit in these pages that the greatest human possibilities rest in a reformation in the way we think and act to direct our technological momentum. We require a new inspiration to match our technological audacities—and to reshape them humanely. No subject offers more varied solutions to the crushing social issues, or more promise of social progress, than the search for new inspirations in cities designed for the grandest as well as the humblest human aspirations. Indeed, I submit, substantial social vision and progress can no longer take place without a powerful reshaping of cities, society's utmost organizer of modern life.

Cities are the essence of modern life and if there is a grave outlook for society, for survival of life as we know it, that gravity exists most centrally in the form and function of cities. The essence of survival and recovery therefore also exists in cities. Yet cities contain within them the greatest potential for continued human progress. And the heart of an urban form suitable for recovery and progress can best arise in a new, dynamic, multi-faceted structure of urban community. My aim is to demonstrate that potential.

Already in these first lines, one may see that I have a deeply pessimistic view of cities overwhelmed by a boiling optimism regarding their promise. And so it is throughout this volume. My hope is that it will strike a match in a furnace, and then forge ever-greater ideas of living creatively in cities.

The vast creative wealth inherent in cities is, I believe, the missing force of modern progress, potentially able to set forth many dimensions of social and cultural development, converting overbuilt transportation into environmental harmonies, establishing living efficiencies possibly more significant than those of industry, giving direction and balance to economic growth, and setting us on a course to a new dynamic of life and ecological validity. I attempt to articulate these possibilities and to formulate concepts that can set cities on a new, positive, and entirely human line of development. The inhuman development of cities to date has left them as merely as receptacles of development, not active and directive forces integrating the content of human progress.

We generally think of cities today as neutral, almost nonentities among the social behaviors existing within them. But cities are not warehouses of industry, commerce, and housing, as if working, buying, and living constitute aspired qualities of life. To the contrary, we might argue that cities of the modern world *are* society with a growing dynamism since the ancient Greeks began defining the good life. When one thinks of urban form very deeply, one is compelled, it seems to me, to contemplate the human potential. Cities can organize good living, human freedom, and ecological validity. However, the dynamics of mind today seems to avoid thoughts of joining urban form and human potential. So I think that until social leadership brings the vast human possibilities into clear focus, urban form will remain on the demented side of change. Such lack of deep concern for this paired relationship of cities and the human possibility has long been destructive and is now becoming deadly for society.

This work is a hypothesis of urban form and function, emphasizing physical design and public institutions, and attempts to expose the huge, specifically human possibilities now latent in society, especially by the creation of community. The city is then conceived at once as an unused instrument of *social* efficiency, equally as an instrument transforming wealth into a great new span of human opportunity, and as a symbiotic union of the built and natural environments of life. The inherently complex power of the city can therefore be shaped to build a new social health at many levels, open a bright new horizon for democracy, and address fundamental ecological questions in surprisingly positive terms. I describe the wide combination of these possibilities as a "grand harmony." But exposure to this round of concepts does not establish anything like our current customary visions of the future. The reader should therefore be ready to confront a challenge to both a very different concept of the future and a new means of working toward that end.

My hypothesis of an urban harmony includes a dimension of beauty. Today, people struggle, spend their resources, and garner their time to escape their ugly, barren, and dangerous urban environments. A plea thus stands behind the hypothesis: that we challenge ourselves to *create and appreciate beauty in our home cities*—indoors, outdoors, and with ourselves and our associations. Then we may learn the contagious character of all kinds of beauty. And only then will the pull of natural beauty, not the push of ugly cities, motivate us to seek a greater harmony around us. Simultaneously, we will elevate our ability to appreciate all forms of beauty, and then stop ravaging the countryside, shore, and mountain.

Ultimately we can learn to build our life experiences by giving them all an aesthetic dimension. But to do so, a fundamental urban harmony is imperative. Such harmony can arise only upon the creation of a new, spirited public ethic. Any useful urban harmony really translates into a broader, deeper ethical relationship between the person and her or his immediate society.

Therefore, the hypothesis encompassing urban efficiency, human opportunity, and a symbiotic ecological union in cities does indeed create a formula that may translate the raw wealth of our industrial genius into a precious human wealth of social expansiveness. For each of us, life is unified. But the legacy of our radically new industrial and urban order is a fracturing of human behavior and experience. To recapture the human balance I believe it is vital to think of new beauties, harmonies, and ethics—which are all made possible, yet deeply disrupted, by our industrial enterprise.

This book is an extended essay and is neither technical nor academic. Nor is it a research effort documenting the cities' many ills. Those ills are reviewed in my books, *Autokind vs. Mankind* and *On the Nature of Cities*, both in this dialogue series. The early chapters herein do, however, review conditions of cities that have a direct bearing on the concepts presented in the later chapters. Another book, *American Communities*, reviews communities in a broader philosophic perspective.

As an extended essay, this book is intended to provoke and stimulate new lines of thought, not to prove them. In presenting a number of ideas, my theme cuts across fields normally dealt with as specialties. Indeed, bringing numerous fields into one vantage point may itself be considered as a central concept. Those ideas are set forth as human opportunities, all together as integrative spaces and facilities. Since our cities today are in such disarray, I found it imperative to emphasize the city's integrative potential, which is really, we must admit, what cities should be all about.

The concepts presented here, I believe, are not difficult to understand. But, being uncommon, they may have a gravitational swing that people are not accustomed to, especially for those who are steeped in the purely economic or technical definitions of progress. Doubt is therefore expected. I do hope, however, that the particular ideas and their sum total will pique the reader to reexamine some precepts that our fast-changing society has, nevertheless, steadfastly implanted, such as building a high standard of living. Communication, especially in the realm of ideas, is always difficult, for this is where we, both writer and reader, mix our separate and complex heritage of thought, and carry our greatest luggage of values. Both our values and logic are thus mutually infused, and the outcome of communication is unpredictable. The best outcome one may expect, then, is stimulation for a wider range of thought and public dialogue.

My training and experience as a city planner helped set me on the track leading to my present understanding of cities. But it also put my mind into a clutch of limitations that I now consider misdirected, many years after leaving the field as an occupation. More than ever, I believe, new ideas require a cleansing or neutralizing of old ideas, creating at least an openness based on achieving an ability to profoundly criticize and evaluate the way things are. Notwithstanding good intentions, urban planning tries to implant modern technologies into nineteenth century urban forms without recognizing the greater human possibilities, without creating urban forms with higher physical, economic, social, and cultural validity. City planning today is *physical* planning, and that confinement robs it of broader social purpose and heart, as well as complementary methodologies. It consistently retreats in the face of the forces destroying the city, the larger framework of human progress, even as it maintains the appearance of fulfilling the public interest.

Similarly, my background in sociology was instructive—while also frustrating. Since I personally seek ideas as a guide for action, the overwhelmingly passive, even indifferent role of sociology baffled me and seemed to leave concepts half formed, awaiting a fuller, more effective development in service of society. Academic ideas today in the social sciences, it seems, do not easily evolve toward applicable ideas comparable to those of physics, geology, or biology. That condition also reflects the fact that applied social science is a limited field, confined almost entirely to resolving problems such as crime or social welfare, not pursuing a human vision.

My approach today is that of a generalist (see the discussion in chapter 8), which means that I range freely among as many disciplines as my personal scope allows me and my method permits. That scope, I believe, especially helps me keep focus on the specific human possibilities, even as it leaves me open to much specialized criticism. Certainly, I am free from subordination to special institutional interests. I take particular pride in being free to press against all limits confining the human condition. So, as a generalist, I fly high and wide, where perhaps the oxygen is thin but the view of humanity is grand.

While this book has been projected for twenty years, with detailed outlines made long ago, writing it in the mid 1990s nevertheless took on a life of its own, as writers frequently report. That seemingly independent momentum resulted, I believe, in a greater strength and richness, such as: in analysis, a more vital connection between economics and urban form; in judgment, a more direct sense of the human impact made by cities; in concept, a better understanding of how urban form affects social freedom. Little was really new, but the content seemed to blossom. My wish is that the reader will discover some of that flowering.

The reader will note that there is no chapter on transportation. This fact reflects a significant shift of emphasis characteristic of this book. Rather than perceiving transportation as an endless good, I emphasize the concept of access, which I believe is a more relevant concept for cities. With effective urban design, urban access can be radically improved without being dominated by the endless pressure to expand urban movement. Poor urban design emphasizes movement, which explodes the demand for wasteful uses of land and, through urban sprawl, sets into motion a radical deterioration of urban efficiency.

My review of the urban condition and urban potential over many years and during the writing led me constantly into the realm of economics, almost to the point of its becoming a major theme of the book. I am not an economist and do not enter into technical or academic economics. My concern is restricted to the outcomes and consequences of economics, especially urban and social outcomes. That concern is central in another book in the dialogue series, *The Runaway Economy.* In a democracy within a very dynamic age, I do feel strongly that economic affairs are too important to be left to the economists. In any case, economics is ultimately a *social* matter of immense public importance.

I must emphasize that this book is not presented as a panacea or a utopia, which to me are largely meaningless, but rather as varied demonstrations pointing to the vast, if unknown potential, each concept to be considered on its own terms but within a broad philosophic context. This potential speaks to the human quest represented in every peak of civilization. Our own quest today represents the greatest step ever in the long march of all peoples to a higher level of human aspiration. That aspiration is deeply a part of us, and to divert it seems almost to destroy it and ourselves in a simultaneous creation-destruction exercise by the independent self-propelling forces of the techno-economic system.

For me, a vision of the human possibility in cities is far more important than any specific concepts, principles, or designs. With vision, many particular options will appear, one way or another. We can see, for example, how science succeeded because there was a growing popular understanding and support for it, especially over the last hundred years. Indeed, today we are swamped by human promise, yet also by progress that gluts our lives, it seems, as much as it fulfills our needs and may widen our aspirations. If a vision is established in people's minds, then the specific developmental concepts outlined here can serve as elements of the public dialogue.

My happiest realization occurred when my method as a generalist revealed numerous harmonies among the physical, economic, ecological, and cultural principles. When articulated, my notes of the last two decades said so much more together than they ever could have apart, and radiated new meanings and concepts, culminating in the grand harmony of urban principles. These positive reinforcements among seemingly disparate realms of human interest reinforced my faith in the high potential of the generalist as a separate discipline capable of enlarging the social effectiveness of our endless array of specializations.

Conversely, following the conclusion of the Cold War, it seems that we also must demobilize from another cold war, a kind of internecine war now deeply imbedded right in our urban environments. That undeclared war builds urban friction and destruction as a virtual principle of development, creating cars and freeways, for example, in ever-greater numbers to overcome the ever-growing urban contradictions of cars and freeways. Ultimately, this grisly war has resulted in a physical devastation in the city and human consequences as vast as many hot wars.

I dream of the city because I find no other vessel of society capable of encompassing my dream. I dream of the future, and the future is inevitably urban. Yet even as I dream of inspired cities, I recognize how far removed my thought is from customary practice. With that dream, I also recognize acutely the huge rubbish heap of discarded ideas of thinkers throughout history, ranging from Plato to Whitehead, or the utopians, and to the greatest trial ever given a single idea in history, Marxism. Nevertheless humankind lives by its ideas. Its greatest ideas of our age are science and technology combined with business and industry, and there are so many ideas in this branch of creativity that they powerfully congest our lives—especially our cities so tragically malformed. So, of necessity, humankind must learn to apply ideas to ideas, that is, to give thorough and reasoned direction to the uses of inventive products and techniques. Inevitably, again, this means the city, for it is our prime social integrator. On that note, therefore, I put my *card* on the table and will be pleased if it can help stimulate a profound debate about the organization of life in cities. But let no one doubt the importance of such a dialogue. Our very humanity is at stake. Indeed, a new higher humanity also stands before us. So, again, my pessimism and optimism are really one.

If I stand perplexed at any single matter, it is this: How does our society—otherwise so rational, scientific, purposeful, and progressive—act to the contrary, and be so irrational, unscientific, confused, and regressive on matters pertaining to urban form and function? Cities seem to exist outside of the tests of modern accountability, whether economics, organization and management, unified and integrative design, or reasoned use of environments and resources.

I believe that most of humanity's problems of modernity—and of the struggle to become modern, as well—rest in some significant way in cities. Certainly our gravest questions of ecology center on cities. Many sectors of the economy promote raw wastage, then profit from the economic stimulation created by the waste. Social disintegration could easily be called wasteful urban disorganization. Education, now locked in barren isolation from the city, could be seething with energy in every part of the city. Instead of healthy living, today's city promotes accidents, violence, drugs, diseases, and many disturbances of the mind.

It was not difficult for me, therefore, to conclude that the city is now organized to be an environment suitable almost exclusively for economic exploitation. That exploitation, it seems, is built solidly into urban anatomy and physiology and has also, therefore, made our cultural genes of social behavior more addictive than rational, creating a tradition of mutual exploitation among us all. Have the words *do unto others* now come to mean that service to one another is edging ever closer to an exploitation of one another?

My perplexity approaches absolute astonishment at what appears to me to be a cultural blindness to the environments closest to us, those we build, where our lives are all but universally played out. Inexplicably, the visions, insights, skills, and momentum of our whole society seem to be simultaneously brilliant and blind, as we sometimes see in the contradictions of individuals. But the universality and depth of our social blindness to our habitats and to our very sociability is profoundly dismaying, especially when viewed against the vast range of society's inventive genius.

Throughout the text, many section heads in the chapters include quotations. These break the pace to establish a point of departure for the section. Thoughts standing alone often demand a level of attention that is unlikely if woven into the text. Consequently, many of the quotations used here take on the character of aphorisms.

I ask to be excused for frequently using the present tense in expressing characteristics of cities that I propose. The style is less argumentative and easier reading than one constantly using such terms as *should, must, or recommend.*

On a personal note, in a less happy day society formed my mind as a warrior to defend our way of life, and I took to it with a sober determination. There was always a seed of another kind within me, however. While I recognized how human greatness and courage could appear when societies collide in brutal destruction, I came to appreciate that the real place for honor, courage, loyalty, and commitment is at home in positive pursuits. Consequently, I asked myself many times with youthful idealism and naiveté, "Why can't we build peace like we wage war?"

Now I know that my motives then and now are the same. And if I may give myself credit, I take pride in carrying those motives from a time of large-scale, organized inhumanity and have struggled to find every possibility for organized humanity to grow. I now take heart in the prospect that the weight of history may have shifted toward the grander possibility. Broader visions of humanity now have the possibility to be formulated and take command upon their merits.

Our time demands visions capable of encompassing the two great opportunities of our time: to build permanent world peace, and to make every particle of change human. The particles of change are what concern us here.

A number of people have reviewed the manuscript or otherwise assisted me, although the result is mine alone. Many thanks are due to the late Russell Fey, the late Roy Potter, Calvin Simons, and Randy Wells. Rodney Nelson helped me with language. My greatest debt is to my daughter, Leslie Schneider, who helped me through the treacherous computer rapids, but where at numerous times I was rescued by my other children, Lowell, Mari, Loren, and Matthew. All assisted in other ways, always in good grace. I wish to deeply thank the following for help they have given to me over the years: the late Ralph Kelly, my sister, Doris Delgado, the late Marilyn Kelly, Ted and Peggy Larsen, Bonnie and the late Wallace De Pew, and Arthur Everett.

PROLOGUE

ON HUMAN POSSIBILITIES

There is a common assumption, mostly unspoken to be sure, that a high standard of living must be a heavy burden upon ecology. There is another mostly unspoken assumption that the high standard of living we now enjoy must command unlimited human energy for unlimited growth of the economy. These two assumptions rest upon a third, namely that organizing our lives for good living must center upon endlessly expanding consumption, focused most centrally upon a large suburban house and multiple family cars. However, a human tragedy arises from these three dominating assumptions, for they constitute a modern totalitarian regime approaching tyranny.

Those beliefs constitute a basic historic mistake of development and this book presents a frontal challenge to that destructive order fixed into the structure of national behavior. Once a plenitude of material wealth exists, this structure of monolithic economic determination represents numerous aggravated economic assaults on the good life to continually expand production and consumption.

This structure of economic determination impacts itself destructively upon the form of cities, the essence of democracy, and ecological viability, together our best anchors of good living, no less than the basic role of economics. This book centers upon cities, a subject no less critical than basic economic production. But cities lack the theory, organization, and defenders that corporate economics enjoys. Cities are the critical environments of modern existence, but are now so critically organized as merely another immense economic market.

The greatest tragedy is that the urban, social, political, and ecological disharmonies establish new realms for immense economic growth.

The assumptions behind the dominating concepts of production, consumption, and their human commandments at the foundation of our modernity has numerous penetrating roots: resting upon the distractive avalanche of materialism; derived from the most energetic system of autonomous enterprise; supported by the most prodigious self-generating system of amassing capital; founded upon the most radical innovations of technology; utilizing the most successful development of bureaucracy; based upon the amazing scope of scientific

1

discovery; and resulting in history's first nearly universal urbanized society. Little wonder, then, that our society is exploitive of nature, socially unbalanced, single minded in its goal, and blind to the deeper consequences of its massive historic thrust.

Culturally, our society is thunderstruck by its own dynamism. Having evolved with seeming ideological virtue and unmatched material progress, free enterprise penetrates politics and social affairs to become a monopoly of social transformation and thereby rules the emerging conditions of human behavior. Its monopoly is propelled as the greatest domestic game of power our society has generated. This game is ostensibly a pursuit of profits; but in reality it is focused on the most modern development of power, the game of controlling the transformation of society.

Yet, we know, and should now know more than ever, that all power is dangerous, and that the modern economic game of power defines us to the quick of our spirit. Those conditions mask the power that also is able to define the entire process as social benefit, while protecting the giant game of power. Still, that game reveals an immense hole within our vaunted processes of reasoning, one that avoids even applying the rationalities of business and economics to the society as a whole, including also the logic of sophisticated technologies or the methodologies of science. Where these rational systems of thought might apply to some larger social issues, they are withheld and denied in applications to the entire society. That itself is a measure of the massive modern powers surrounding today's economy.

Altogether, the economic powers applied to business decisions are not allowed application to the broad directive forces of radical change. They reflect the conflict arising from the environmental movement's struggle to apply those rationalities to the entire society. The inconsistencies arise from continuing to apply the historic rationality of a time of harsh scarcity to the entire thrust in this time of wasteful abundance.

As a consequence we confront a bare-faced imperative to terminate the ecological destruction and another imperative to halt the social disintegration, with both now imposing huge deprivations upon society, as if a new economic scarcity will be necessary to avoid ecological and social catastrophe. But from the concepts presented in this book, it is not only possible to resolve the two imperatives head on, as simple pragmatism seemingly requires. Society can make major adjustments to face its crises while simultaneously inspiring and underwriting a very human renaissance, a progress focused very specifically upon the well-being of every person. In other words the two acute tragedies of modernity can be corrected as part of a new course of positive social progress. By applying the rationalities found in science, technology, and in economic enterprises themselves, both the dire imperatives of our time, and the huge human potential can be pursued together through a combined rationality and a newly inspired social ethos.

The great myths we must confront arose out of the Industrial Revolution. The reasoning was that the single issue of society was to build more product and eliminate the acute scarcity faced by humanity through the ages. It failed to consider that organizing consumption for the people, that is, organizing society specifically for good living, held as great or greater human potential as continuing to expand production. But as productivity reached affluence, the corporate economic leadership discovered that it could continue to increase production and profits by simultaneously continuing to expand raw consumption, whatever the consequences. The greater human potential was ignored in favor of unlimited consumption. Meanwhile the raw economic growth continued to promote deeper ecological and social issues we face today. My purpose is to disprove not only these tragically mistaken beliefs but, more importantly, to set forth a different logic of development by which progress will be understood as a harmony with nature, not the multiplying consumption of our earthly resources.

The key to good living, once basic human needs are filled, exists primarily in cities. But cities, it turns out, became the environment that could most directly be governed to command greater consumption. *But urban efficiency, rather than increasing, steadily dropped as people abandoned the old inner cities, used their new automobiles to access the countryside, and compelled government to reconstruct the cities as machines for moving rather than as environments to live.* The more inefficient cities became, the more they demanded enlarged economic output; so inefficient cities today contribute more to the GDP than efficient cities. That measure in reality reflected ecological and social destruction, not a growth of human benefit. Hence, we have inherited the ecological and social destruction of misdirected economics.

In truth, as I argue, a high level of material consumption directly impedes a higher and more generous quality of life, and is ecologically destructive, and congests the processes of democracy. My thesis is reasoned testimony that the human potential of cities, democracy, and ecology is greatest when each is in harmony with the others. The essence of good living is an unburdened plenitude when ample but restrained consumption actually frees us—our money, space, time, and emotions, the entirety of hour human potential—for the fullest possibilities of our lives.

In the short history of the twentieth century, in little more than one lifespan, an immense human promise has been created, yet largely denied by the same forces that underwrote it. Both the creation and the denial are consequences of the technology-and-profit-driven economy and, ironically embedded in the wealth-making process itself. In possibly the greatest paradox of modern development of society, what is made possible is denied.

The true promise of the material advance of our time is not a high standard of living, as happy an historic event as that once was, but rather a broad range of fundamental and essentially new freedoms, that is, social opportunities that could have progressed along with the material advance. But beyond a limited number of conveniences and the modern span of canned entertainment only a limited social advancement has appeared, mostly that which stems from mechanical and consumer recreation products, such as scuba, sky diving, and various racing equipment. Instead, we live harried lives cluttered by the force-fed consumerism. If the economy has created a historically wondrous array of instruments to advance our abilities for living, the same forces simultaneously diverted us, blinded us, and organizationally all-but prohibited the greater social advancement.

Obscured by the onrush the new wealth are great human defeats: urban decline and chaos; drugs, crime, and the explosion of prisons; costly schools that teach less to today's students than they did for their grandfathers; and isolated suburban living that promotes television dependency; and a deprivation of children's playmates and free-learning behaviors. But rarely are these losses measured, for example, against the wealth exhibited by costly cars, oversize houses, hardly-used recreation vehicles and equipment. While autos are given credit for uniting people with the city's production, consumption, and association, in reality the car's effects are profoundly isolating, especially when joined with large-lot, single-family houses, and the freeway's scale of ecology-defeating distances. Then, building upon isolation in cars, houses, and most recreational equipment, the social isolation goes into the house itself where both the singly prepared and eaten microwave meals and television viewing convert interpersonal behaviors into psychic dissociation.

Both the obviously powerful and subtle forms of human isolation therefore defeat a truly human renaissance ostensibly underwritten by the vast new wealth. Cities—the potent social incubators of civilization—have lost much of their essence of close, healthy human association. Then, when the environment and social behaviors are structured for isolation, a society of alienated persons is converted to its true condition, socially undifferentiated urban masses susceptible to the raw promotions of advertising-driven mass media and a SuperBowl mass mania of directed behaviors.

The phenomenal twentieth century has seen the greatest race to invent and produce ever known (as well as the greatest wars and inhumanities). Generated at the same time is a gigantic rainbow of forceful mechanical, corporate, financial, and political powers ever to inundate a society. Tools of this historic power serge include vast resources, an explosion of new skills and technical capabilities, an unprecedented mobilization of knowledge, regular innovation, and not least, the sophistication of corporate bureaucracy combined with its relentless initiatives. Vast and varied business capital underwrote hundreds of thousands of business ventures necessary for it all. Government largely stood aside or actively helped ignite these fireballs of promoted change.

The city consequently became a huge mass of activities with a continual loss of its human and social integrity. The rich cultural dynamics of cities themselves could only barely exert itself. And, since the growing wealth after 1945 permitted squandering of resources, the cities themselves became sinkholes of destructive inefficiencies. With the promoted chaos of yearly new models of automobiles, the cities were left in such masses of movement they futilely struggled to accommodate the onslaught by building the transportation vice grip of roadways. Those streets, boulevards, freeways, and the fields of parking today not only dominate all physical exchange but also generate the modern defeat by mass daily commuting by automobile, the least efficient and most consumptive mode of urban transportation ever devised.

With such promotional opportunities, there seems to be no natural limits confining the corporate boards of directors. Only the next hurdle of promotion concerns them. Posing costly problems for society meant merely that there is as much money to be made from the growing problems as there was in promoting the original products. In other words, the completion of a new freeway opens the way for more people to own more automobiles and to drive them greater distances to more isolated suburban districts. So the market demand for automobiles is built in conjunction with the popular demand for more roadways. Thus the follow-up challenges established a runaway economy in which urban and social chaos become as vital to economic growth as the original human wants.

With cities and the entire society cast as numb ingredients of corporate growth, the economy can then be spurred on by degenerate urban conditions. Cities therefore have become ideal environments to promote greater economic demand, that is, by creating new inefficiencies and building entirely new scarcities that promote new economic demand. The runaway economy thus grows upon synthetic creation of its own market demands. This permits economists to emphasize the never-ending "scarcities" and manage the economic levers for the unending growth of the infinitely expandable GDP.

Yet, possibly the more critical long-term consequences of unending growth stem from blinding society to its specifically human future for, among its effects, the runaway economy can continuously stress its own purely market demands at the ironic expense of other specific human possibilities. Thus economic growth becomes juxtaposed against a broad human social and cultural potential founded so paradoxically upon, the economic expansion.

The numbness of cities illustrates a far greater social issue than the commercial chaos in the service of economics. The modern perception of cities by society exhibits a functionally abysmal condition, especially when we consider how we have been blinded to them: the most vital environment of modern life; the central structure defining human ecology; the content of social life; the daily efficiency of business, government, and persons; a learning environment as critical as schools and universities; and certainly, the incubator of culture or crime. We don't take pride in building socially creative cities, as we do in sports, recreation, and music. Instead, we take our pride and escape to the isolation of distant suburbs and then leave older sections of the city to decay, upsetting the historically normal processes of urban evolutionary improvement, sometimes to levels of grandeur. That process applied not only to the Athens and Romes of history, but also to Jerrash, Pompeii, and hundreds of other ancient cities not blessed with modern capacities of construction.

Money is inevitably involved since it is both society's greatest tool for cooperating and sharing and the greatest common motive to achieve individual power. Today that tool grows increasingly upon the promotion of economic, human, and ecological waste. The important question for our age, therefore, is how we can stop the runaway economy and simultaneously build new, higher forms of *social* cooperation. Where are the concepts that will unify the city for the person and give urban form a new force for democracy? How can a reordering of economics embed a new force of rights and equalities squarely into the physical and institutional structure of both the city and society? Can we build cities with efficiency and grace while capturing the technologies that are ecologically valid?

The answers are, I believe, much simpler than the research and development we now pursue with so much discipline and money. After years of thought, I have concluded that the chief hurdle may be a new inspiration that gives us a renewed view of life and a fresh sense of what society is all about. Then perhaps we can create a new meaning of being a person in which social interests rather than economics directs the course of change. Yet, an empowerment of the person will not happen without an equal empowerment of the close interpersonal groups. That is where a profound rethinking of the cities becomes imperative. Cities are where democracy can best be enlightened and expanded, and where human ecology can be given a sound and enduring constitution.

Cities unquestionably must become the center of any modern social renaissance. When we begin to think about them seriously and ideally we will find many new and creative elements to be disarmingly simple. But first we must inspire ourselves, paradoxically at the outset by recognizing the urban pathos. We may start with a simple proposition that is at the heart of so much of today's social and ecological distress. *Cities are now organized to destroy wealth and our humanity so that they can promote an endlessly growing Gross Domestic Product.* Their wastage contributes to the GDP. An auto accident, congestion, a crime, and the multiple consequences of escaping to the suburbs all make an ironic positive contribution to the economy. Proof is no more profound than making some personal observations about the causes and consequences of the urban distress. Then we can see the true significance of the runaway economy and the sadistically mistaken process of measuring progress by money flows. Let us consider seven closely interwoven cases.

First, why do we promote urban transportation as an endless good when what we want is the most efficient form of *access* to what we most need in the city? Access must be achieved through a new, compact design of the cities' built areas, uniting home, work, shopping, recreation, education, and group activities as much as possible. These can be beneficially integrated—as in community—so that little formal transportation is needed at all. And much of that can be accomplished with ease by the most efficient form of movement of all when urban design respects the human scale: our feet. Then additional and formal transportation, like transit, can be built to serve the needs that could not be accomplished by design.

Second, why do we build urban environments that are so prodigiously beyond the scale of our bodies, sensibilities, and behaviors while so wastefully consumptive of land, resources, and nature, and while also demanding so much pavement for so many vehicles to become so congested? Traffic congestion, we must remind ourselves, is a product of dispersed urban development, not compactness, which can bring so much of the city within an easy walk of a few minutes. Do we build the city beyond our human scale simply because we can?

Third, why do we build the city for so many forms of human isolation? Human beings are social beings first and foremost, and our humanity is sharply diminished by being monopolized by utilitarian contacts without the rich interpersonal seasonings. Without many varied and close relationships we too easily become beasts by degree. We could build the places and organizations of the city to cultivate endearing association. Community has been one of humankind's most consistent and personality-building pursuits, and the soul of community continues today, however brutalized by the utilitarian directives of our race to high production.

Fourth, why do we consume so much land, carve it into so many miniscule parcels so little used but requiring so much work? We could take half as many acres, build on but a fraction of that land and then reserve most of those acres for large, varied open spaces, thus giving everyone who wishes a regular, varied, and close involvement in the outdoors. Then the massive, dreary pavements and the unending monotony of private mini-spaces would reveal themselves for what they are: massive attacks on the environment for the ego satisfaction of possession and consumption, aggressive isolation in urban sprawl, and subordination to canned commercialized entertainment.

Fifth, why do we build urban environments that are so grossly and obviously inefficient? And why do we do so when by every social and ecological perspective—including sane economics—the results are disastrous? The only plausible reason is that the marketplace, pushed by corporate profit making, is promoting an eternally growing runaway economy. When money is rationalized as the chief motivation throughout society, it then becomes the supreme condition of all human life.

Sixth, why do we create schools completely in isolation of the many contexts of life in which learning can best occur and in which life must be conducted? Quite simply, we build cities in which the major activities are isolated, closeted out of view, and thus beyond the reasonable reach of education. Today's education has become another industry preparing people to produce and consume to keep the industries active in their runaway mode.

Seventh, is it possible to idealize an urban environment composed of highly efficient human communities that have many of the combined qualities that are equally valid socially, culturally, ecologically, and of course, economically? The thrust of this book is that such a community is eminently possible and might be a concept worthy of becoming a major human ideal in the city. Community can become a solid base for pride, tradition, and a kind of love we have so successfully rid from society.

The significance of all seven observations is the same. Together, and including many other issues to be considered throughout this book, an urban and social renaissance of the most profound order is highly feasible if we could view cities human ideals and create them with imagination.

Cities can complement and integrate diverse human interests for optimum combined development: solid, natural and ecological validity; great business, public, and personal efficiencies; environmental, architectural, and other design arts; economic, political, social, and cultural organization. They also highly complement each other, which justifies the term "grand harmony" in the subtitle of this book. And the harmony most poignantly underwrites a new scope of *social freedoms* achieved by creating an endless number of specifically *human opportunities* in the reorientation of society from the disastrous runaway economy.

The integration and unification of many fields for inspired cities derive from a broad methodology I use as a generalist, a field deserving to become a counterpoint to the sharp focus and great force of specialists who reign alone today. The city and the generalist are uniquely suited for each other, both uniting many purposes, interests, and functions for optimum human outcomes, that is, for outcomes focusing on both the whole person and whole society.

The lack of a very broad perspective that a generalist might promote is reflected in the conservation movement, which has grown into possibly the most powerful citizen movement since 1950. Although the movement ostensibly spans the entire range of ecology, it does not reach into the cities except in today's focus on urban sprawl, despite the fact that cities are the destructive center of environmental decline. Somehow *urban* ecology is neither deeply a concern of the conservationists nor a worthy movement of its own. Ironically, because cities are so destructive in their present form, and an increasing number of families are escaping to the distant suburbs and holiday retreats in the mountains and along shorelines, they promote precisely the same kinds of destruction in the holiday sites as in cities.

The challenge ahead is indeed mountainous. For a new inspiration to awaken us, we first require the new, inspired perception of life in our highly potent society now so dominated by making and consuming. In reality, a new paradigm of society is imperative, one that goes beyond the material inventiveness we have sought so singularly for over two centuries. While both the deadly overhead and the great promise are immanently before us, both economists and corporate leaders still refuse to recognize ecological and social limits. The classic model of free-market expansion continues to prevail, even as evidence of runaway economics becomes overwhelming as ecological destruction and human distress. Yet, so tragically, society continues to stand with the economists who give their moral blessings to those who direct corporate power solely for money.

Since we live in a society dominated by economics that is where we must look for the fundamental powers of today's society. However, the dominating profession of economics was formulated at a time of relatively static conditions of general scarcity. But today the chief clients of economics are the corporations who seek profits by promoting destabilizing change throughout society. So the ironies for both ecology and social existence are that the economics once conceived of a static free market of small buyers and sellers is now applied to a one-sided market by very large and powerful players promoting an unprecedented onslaught of disruptive change. Thus, an antiquated historic economic model based on static conditions with unrelieved scarcities dominated economics of commercially promoted change continues to prevail in which ecology and social life in the old view are now but minor and passive players in the process. *The modern, powerful marketplace simply cannot respect non-economic purposes when that society is organized solely by and for profit-seeking corporations.* The essential lesson for us today is that the market, highly biased for corporations, cannot continue to be the legitimate legislator of change throughout society.

The problem of our fast-changing time is that neither the political system nor the economic processes were conceived or organized to guide the vast social transformation. They merely promoted expansion. Clearly, the GDP vividly dominates our national values. A broad social goal system pursuing the human possibility has been completely preempted. A money-calibrated, non-goal system of evaluating progress as consumption is substituted for varied human objectives. The increasingly urgent matters of ecology and social life remain naked, oppressed, and passive players, quite incidental factors when set against the powerful and deterministic, profit-demanding corporate economy.

The real poverty of our time—stemming from the historically sudden material profusion—is imagination of the human possibility. We are misdirected to amass things and obsessed by the novelties of high-tech. So one day our era may be characterized as the "material enclosure movement." Breaking the enclosure and stimulating a new direction for the social imagination will depend upon two initial foundations: (a) building an awareness of the suppressions connected with corporate-forced materialism and (b) creating rich and powerful public goals for the political economy to underwrite.

If we first set a number of specific human goals and become involved in achieving them, we will discover that new horizons will arise and entirely new outlooks will emerge. Our sights today are not only depressed by overloaded materialism; they are also constricted by the tunnel vision of money and boxed in by the greatest interlocking directorate in history. A penetrating smog dulls our social vision pervasively; it beggars everyone to clear out old images to conceive a life in which economics serves us without more ado than society now produces food. Then, I believe, our minds may be reset to a new paradigm of the human possibility.

As an effective instrument of society, the economy can be challenged to provide the liftoff for a new essence of human experience. This poses the enormous question of what we must ask of the economic establishment and its technical components. Three initial objectives are fundamental. First, elementary justice demands that everyone in society benefit from a foundation of plenitude, the personal security upon which other goals must rely, the security by which people acquire the personal confidence to reach out for new experiences. Second, we must establish an ecological validity, not merely sustainability, but no less, restoration and enhancement. Third, society needs to establish a new scope of free time—the essential ingredient of civilized advance—including that for real leisure and much serendipity, out of which highly varied creativity will emerge.

With a plenitude, ecological harmony, and free time, the remaining social challenges underwritten but not controlled by the economy, are personal, social, and cultural, that is, the realms where greatest imagination can arise. Most pointedly, a major focus may establish a *freedom to be*, a buildup of personal character that enlarges the capacity to experience life in many new ways: physically, emotionally, artistically, intellectually, or by a thousand kinds of adventure, discovery, and love. An early discovery will show how profoundly suppressed our ability to experience life has been. By using methods of self-development and paradoxically by building individuality through community, society can stimulate and nurture a new strength of personality. Personality arises best by a leavening of person-to-person relationships.

Then, upon new individual strengths, *positive social freedoms* may be generated which provide the means for creative endeavors to be conceived, promoted, and fulfilled, either independently or by groups. Here, then, is a new scope for democracy to expand the positive dimensions of life, true positive freedoms, unlike the merely protective guarantees of the Constitution.

If social ambitions spread to large populations, then perhaps society will create vast modern moral equivalents of the Egyptian Pyramids or the Great Wall of China—but upon cultural and democratic foundations. Thus new social challenges and great achievements of society yet to come will rest upon human dynamics. All of us will be astonished by our illimitable social possibilities. The essential objective overall is to build a grand harmony among environmental, social, cultural—and especially economic—fields upon which the human experience can flower without life confining obstructions.

CHAPTER I

PROMISE OF CITIES

Cities Organize Society

> *The city is at once man's greatest achievement and
> his most abysmal failure.*
>
> Harvey Wheeler

We live in the beginning. The beginning of new beginnings. World revolutions of science, economics, and politics. The dawning of the first universal civilization. The rapid urbanization of all life. The greatest beginning of all!

The city is the center, the cradle of civilization, organizer of behavior, focus of ambition and power, director of benefits. But while the city today cradles the enormously generative ideas, products, and arts, the city itself is a bastard shell of discontents, wrought asunder by vast, promoted waste, overproduction, multiple crises, and human tragedy. While society eagerly pursues wealth and thousands of special interests on the one hand, it simply reacts halfheartedly to the escalation of grave issues on the other, having been generated by the primary economic powers in society. Upon ideas and programs that might shape the uncounted urban activities into coherent benefits simply have not been called into existence.

Can society creatively broaden its views to respond to the unprecedented challenges and possibilities? Can we capture the largeness of life made possible by advanced technologies? Can this most auspicious historic opportunity be directed to expand the essence of democracy? Can we find a higher vision for the human sojourn on earth? Can we harmonize cities and ecology? Can we build cities as social and environmental jewels of higher human aspiration?

The case presented in this book is that cities are the key to it all.

If new reaches of democracy and richness of life promised by diverse social creativity can ignite excitement for new beginnings, only the span and organizational force of the city is capable of bringing civilization from unfocused grasping to a cultivated flowering maturity. New beginnings now before us challenge the very foundations of human existence, the nature of the human career, the basis of human association, the significance of property, even our very way of thinking about life.

Daily we feel the vast prize of our modernity, the phenomenal creation of exotic new products in the last century, from jet travel and television to computers and compact discs, from fiber optics and lasers to antibiotics and genetic engineering. Our prize is nothing less than the immense increase of industrial output, an advance as amazing as the new products themselves.

But these prizes are not merely a grand inventiveness put into action by mammoth investments and sophisticated management. Just as much, they can be the gifts of cities. Creativity, capital, organization, and skills would be nothing without formidable concentrations of people, articulate combinations of tangible and intangible resources; instant transfers of designs, talents, money, and material. All of these complexities upon complexities are possible because they are articulated in cities, a passive condition so natural that we think little about this side of change.

Yet another greater prize of cities is little imagined. That prize, in time, may be no less rewarding than the revolution of industry, a prize in which the city itself becomes the central, active framework for human progress. We can foresee immense advances in the human environment, paralleling the advances in product. At once, urban designs can enrich interpersonal life with a depth and variety of social and cultural involvement we can now scarcely imagine. People will be magnetically drawn into vibrantly active plazas, cultural centers, and diverse outdoor activities alive with changing human interests, but only moments from everyone's door. We will then rediscover the dynamics of public serendipity and a renewed human scale in which we can rejoice in the voices and eyes of friends, and new spontaneity in the casual way of social interaction. We will also discover quite paradoxically how the more urbanely we build, the more we can bring the richness of nature into our lives.

Just as powerfully, new urban environments can bring great efficiencies into our work, our errands of life, and our leisure pursuits. To date, ironically, while industrial efficiency has progressed dramatically, these very actions have systematically destroyed what efficiencies both cities and individuals once had. Now, however, cities can also develop great efficiencies of their own at all levels of operation, from utilities and transportation to new savings in industry and government, altogether creating an ease of function and behavior now difficult to perceive.

Cities, therefore, can develop their own equivalence of product inventiveness and great productive output. Yet this great gift of cities will not require a new realm of science and technology for realization. Present technologies are quite sufficient. What is required—and in the end what may prove as challenging as new technological systems—is a new direction of thought, new winds of imagination, to make cities into active and positive forces for human progress. The essence of this transformation is redefining and translating technical and economic wealth into specific *human* progress.

Simply put, cities organize society. That power, put to positive use no less than the power of machines, whether utilities or automobiles, can organize machines for much greater service to humanity. Now, however, we haphazardly concern ourselves with cities. I will show how only cities can optimize the benefits of our industrial wealth. What we require is a constitution for cities that creates a popular vision and sets a course of public action that ultimately will underwrite a new range of positive human freedoms fully equal to the unprecedented technical sophistication of our age.

Cities are now the fulcrum between old actions and new beginnings. Their ideal form is very close to the end purposes of society, potentially offering a dynamic of life that we have missed in the singular pursuit of production and consumption. We have nearly lost sight of these higher social goals at a time in history, ironically, when they are most achievable. But the world today is exclusively a product-oriented system of public action. This means that human end values are muted, fractured, or lost, overrun with products that demand more products and services to overcome the urban wreckage and despoliation of the environment, yet while smothering so many pursuits of the good life. Automobility is self-generating and constantly demands more roadways and parking, and accesses vast land for urban sprawl, which then demands more automobiles. While mobility dominates the form of cities, then crime and police, neuroses and therapy, all become part of the snowballing demand in ever more tragedy-laden cities.

The multiple crises of education, health, crime, trash, and degradation of the environment reflect the historic emergence of massive productivity set against the absence of a guiding purpose to make all output completely serve human ends. The human tragedy arises in the disparity between overwhelming productivity and the impoverishment of methods to make all output and all services fully effective and beneficial.

If social ends of society are ever to shape the means of life in society, urban environments and institutions must become the chief arena to create such an ends-means harmony. Not to do so is a blemish upon human intelligence and blights democracy.

The urban environment itself is likely to become a critical—if not the best—measure of the good life, leaving per capita income increasingly peripheral, just as food production has become peripheral to society. The prospects are not merely a cleaner, better landscape but also an enriched urbanity of great human opportunity and an entirely new dimension of freedom—in cities aspiring to become modern Athens that will give spark to personality and fire to human behavior.

For two centuries, creativity has centered upon science and technical invention on the one side and enterprise bureaucracy on the other, relegating most other forms of creativity to the backyard of society. However, historically speaking, our radically new and complex social circumstances offer immense opportunity but demand creativity in new arenas. It is our task to reconcile our great creativity and our ever-expanding human possibilities.

American society, unfortunately, has made technical invention virtually into a tragedy. As a result, we see this in economic and ecological conflict, in bizarre contrasts of great new wealth set against rapidly declining old neighborhoods, in public subordination to the corporate will, in immense public budgetary commitments but tax shortfalls, and in the highest rates of crime and punishment in the western world.

America's two great historic achievements were settlement of the frontier and industrialization. Whereas the frontier merely vanished, industry continues to grow upon increasing public inefficiency and widespread social distress. Hence internecine output shows no limit and continues to feed upon its own excesses. For example, the pursuit of automobility has created disastrous urban sprawl and dispersed the vital urban core, all while making much of the rest of the city unlivable, all while garnering in its service more taxes, more corporate capital, and more personal wealth to propel demands for ever-increasing movement. Unfortunately urban ideals were never among American ambitions, and could never, therefore, ameliorate the effects of unbridled industrial momentum.

The challenge is everywhere about us. So we must ask if the pioneering spirit is now dead in America. The achievements and daring of America are vast—once the country is challenged and accepts the challenge. We pioneered a continent with men and women that did truly match our mountains. We pioneered democracy, now the horizon of all peoples. We pioneered a hundred spheres of invention and production. We pioneered the trail to the moon with technological pizzazz. Is it not now time, we must ask, to make technology indisputably human, to build social validity into our democracy, to preserve the continent for permanent habitation? Is it not time, therefore, to create the inspired cities that are necessary for it all?

Quite plainly, a whole epoch of history has run its course, and a new challenge with a new outlook, new pioneering, and new, more human achievements stand before us. My thesis is that only the city is worthy of becoming a new central goal of society, for it encompasses the span of modern life. Like the earlier ambitions of the frontier, much of the excitement lies in the pursuit itself. But the city, unlike industry, closely unites the means and the ends of development. The good and humane city may be the greatest objective of development, giving the greatest satisfaction to its residents on a hundred fronts.

Society's commitment to an urban future is inevitable. There can be no alternative. Today, however, there is no social commitment to create fully human cities, and urban development remains dominated by the false race to increase production and consumption, or their unlimited economic growth. All other considerations sadly remain subordinate to that singular force.

Revolutionary We Are

> *Indeed, we find no other animal species that has been as savagely destructive as humankind. In moral terms civilization is something that has not yet existed...might humankind perhaps build the first civilized human society?*
>
> Bertram M. Gross

This essay is a plea and a proposal to bring cities to center stage of our revolutionary society. While the naked force of our revolution is technological and economic, the ameliorative conditioning of the revolution is urban and social. While the technological force today is dynamic, even aggressively exploitive, the urban condition, which can be completely harmonious, remains inactive, even recessive. Both the aggressive and recessive forces at the heart of the modern revolution shape our thought, our expectations, our sense of social purpose, our directions of creativity. They both plead for new vision and a new direction.

If cities are to ameliorate the largely positive but often very wild results of this revolution, that is, to make the revolution an uncompromising benefit to all people—as I contend only cities can do—then they also must become revolutionary. Theirs will be a catch-up revolution but a revolution nevertheless.

As radical as our times are, and with all the gadgets and transformations that surround us all, a serious deficiency has appeared in the logic of being modern. This deficiency is not merely that we as a society have not had time to digest all of the novelties put before us. Digestion also implies selection and preparation of what we purchase or consume. But today we do not organize for appropriate integration or consumption. So now we must ask whether better living can continue to be found principally in continued straight-line acceleration of crude output and forced consumption.

A new logic of modernity needs to articulate all that happens in society to assure expansion of human opportunities and freedoms. These grow only when the physical and institutional structure in society is shaped precisely and completely to expand them. Since cities pervade virtually all that happens in society, it is within them that we must organize to achieve greater social freedom. That is the condition for a fresh logic of modernity. Cities must not only move to center stage. They also must figure centrally in and raise social ideals and aspirations.

Therefore, this essay challenges us to think anew about cities, to ask what cities are really about, and to examine what cities should do for us in these most promising times.

In the United States, the great transformation from exploration and settlement to science and industry was encompassed within newly developed cities. Now we can transform progress into a thoroughgoing social fruition within cities. Old concepts of the city may then pass on like old concepts of production. When that process of thought has begun, we may next look for an entirely new range of human vision. Might we then pursue the development of cities like we pursued product expansion or the development of medicines and exploration of space?

Pivot Time of History

> It seems beyond dispute that the present orientation of society must change.
>
> Robert Heilbroner

A new human dialogue is imperative. It must be directed to all leaders of society, both political and economic, telling them that the wealth of society they helped produce is spoiling in their hands. They cannot proclaim an endlessly growing horn of plenty when the human consequences of plenty are result in consumptive chaos—which ironically includes a large population still suffering from want.

The question now is to learn how to shift social focus from the singular objective of merely producing wealth to emphasizing the better uses of wealth. This will mean a re-creativity throughout society in which the city must be the heart of that re-creativity and the active force behind a new definition of progress. Product affords but does not organize opportunity, underwrites but does not assure new freedom. The city inevitably organizes the form and content of both opportunity and freedom, whether confining or expansive, and so it deserves our most critical attention.

Hope for cities will be possible only if there is a universal inspiration and a solid determination to make them not only a monument to a new and as yet undefined form of civilization, but also to make them a continuous celebration of all that society can bestow upon every person. No one can doubt the technological capacity of world society to do this, even if we now doubt the political will. To be sure, there may always be recessions of uncertainty, misdirection, bungling, political vanity, and state terror. But once an inspiration has spread, the progression to higher achievements is assured.

World society cannot long ignore that this is a crucial time of human history. Magnificent industry is in our hands. Sophisticated economics is at our service. Only cities—the masterful instruments that can shape industry and economics completely for people and society—must yet be seized creatively.

The prize of all human history stands before us. We live in the most auspicious pivot time of civilization. But at this pinpoint of history we also confront imminent ecological and social catastrophes.

At no other time in history has society confronted better and yet more urgent challenges to the very dynamics of change. We therefore live in a true crisis time of history—a vital turning point—in which both the promising and ominous prospects confront us as one. So we must ask and act upon the profound questions of what we are to be as a society, what we should become, and how we should face this crisis of our history.

At no other time has society become more powerful and complex, commanding us to give strategic guidance to the radical forces unleashed in the last century. At no other time have people's lives become more unified and interdependent, which therefore requires finding a common course to follow, especially to develop personal psychological security and freedom. Constructive, unified, and strategic public action must be the fundamental response of society. Incredible it is, therefore, that we should spend so much time scientifically analyzing the social chaos of change and charting the shoals of human defeat, when we might be designing a broad directive to give general guidance that comprehensively measures up to our radical conditions. These options stand starkly before us as a society and can be ignored only at our common peril.

Yet the world's determination to develop through science, technology, and modern bureaucracy has left society without a perception of the vital role of the city, possibly the central element of a broad, positive development policy for post-industrial society. Society has yet to understand that since cities are central to virtually every facet of the modern human enterprise on this earth, and are possibly today's most critical lapse of development. *The current urban condition has become the first tragedy of world development.* Can this gross world misdevelopment be long tolerated? No issue is more central to the human future! Cities are in my judgment the most effective and comprehensive means for development of the entire society. Simultaneously, they encompass nearly the span of society and define most of the desirable ends of development. If progress is to retain any human reality at all, the city's gigantic facts, discrepancies, and failures must be set into the most urgent arenas of public concern.

The city determines the basic efficiency of society, defining what kind of transportation we must have and what products we can do as well with less, or better, without. Urban efficiency defines our basic human ecology on earth. As society's chief producer and consumer, the modern city must become the master mechanism of economic and social development. The sheer size, power, and complexity of cities in hi-tech society make their fate fundamental, which gives them a priority in the highest reaches of thought and governance.

Cities define our humanity. Given imagination, the city will become possibly the most positive player in the future of human development. The question now is how we can conceive positive human achievements for them. They have always been the vigorous centers and inspiration for civilization, as well as the epicenters of power. Historically, they often oppressed people almost in proportion to the centralization of power. Now a new kind of urban oppression bears down upon us, this one arising out of the very form we have given to cities. Society has yet to perceive that its crisis of misdevelopment arises from a gigantic misconception— or non-conception—of the most central and fundamental role of urban environments and institutions.

Yet cities, like governments, can alter their historic roles. We know, in America as elsewhere, that we have yet to build a truly democratic city, one that in structure, function, and social organization gives all people all possible opportunities for personal development, giving people an exaltation in their being.

At this most significant pivot time of history, a new, magnificent role for cities stands before us if imagination and inspiration can be ours.

New, Magnificent Role

> *It is not without regrets that I have come to the realization that invention, in the sense of gadgeteering, must come to an end. But the inventive spirit must not perish; it is much too important part of the creativeness of man. It must now be redirected, from "hardware" inventions towards social inventions.*
>
> Dennis Gabor

The world reached an unprecedented position in early 1990s, one that may well herald the first full flourishing of all humankind. First, the Cold War of nearly five decades came to an end peaceably and overwhelmingly in the direction of democracy, except for numerous aftershocks of international ethnic conflict once held taut in Cold-War discipline. Second, unlike in the devastation of World War II, vast capital resources and new technological capacities now exist worldwide. Third, technical, educational, and administrative capabilities are now becoming fully available to propagate the first universal civilization of humankind in all world regions. Fourth, whereas all previous civilizations on earth materially benefited but a few extremely small, powerful elites and in the United States our own wealth substantially benefits hardly more than a majority of the population, the horizon of a more equitable democratic system throughout the world can promise an immense progress for all peoples.

Here, now in our time, we can see an unprecedented convergence of conditions establishing this most promising pivot time in history: Civilization is poised at the threshold of its greatest breakthrough to a new humanism on earth, one based on material wealth but going far beyond it.

We cannot allow blindness to the crucial role of cities to divert attention from this prize of achieving a new, higher plane for civilization. This challenge and opportunity must become universally recognized, commonly sought, and firmly imbedded in public action if we are to see continuous progress and fullest achievement.

Here is the magnificent prize of history! We cannot permit this promise to be broken or wasted away!

1. The first precondition is clearly a world order progressing toward the abolition of war, especially a worldwide conflagration. Powerful ligaments restraining warlike tendencies also are appearing through national economic development, international trade, world tourism, and international higher education. Growing international cooperation on environments and civil liberties, for example, strengthen those ligaments as well.

2. The second precondition is the growth of worldwide economic and social development. Uneven to be sure, it is nevertheless forceful. The growth will be more promising if resources devoted to arms can be reduced and population growth can be brought under control. The promise of development is that each society becomes free to progressively broaden human objectives beyond raw materialism.

3. The third precondition for a breakthrough to an entirely new and inspiring definition of humanism is the astonishingly rapid progress of world democracy and the rising international pressures for civil rights in each country. From a foothold in only a few countries in 1940, democracy was firmly established in Germany, Italy, and Japan after World War II. Later it expanded to Spain, Portugal, many Latin American countries, and some in Asia and Africa. Time is

required for consolidation in each case, though the general prospects are good. The freeing of Eastern Europe and the breakup of the Soviet Union greatly expanded the prospects for democracy worldwide.

The three preconditions are mutually reinforcing and interwoven, of course. Declining arms, progressive development, and the rise of democracy, taken together, are a completely new phenomenon on earth. Their success is not assured, but their prospects are immensely encouraging.

The breakthrough in sight is a multidimensional freedom, one might say an exponential power of freedom. It is an open-ended social freedom based on expansion of positive opportunities that span the range of known and unknown human aspirations.

One might suppose that this is what we have been seeking all along through economic growth. I would argue, and this book is such an argument, that a singular stress on economic growth has led us to a false promise. Material affluence itself is merely a foundation, like all economic advancement, and is itself not anything like the fulfillment of a higher human promise.

Political freedom, we know, can exist only within a structure of freedom established in law under a liberal constitution. The same is true with a broader social freedom, based on an expansion of positive human opportunities. They, too, require a structure, but this constitutional structure is not merely protective. Rather, it is affirmative, creating the widest possible array of human opportunities.

The principal structure for affirmative human opportunities is, of course, the city. The new dimension of humanism must therefore stem from an urban constitution that underlies, organizes, and actively promotes a greater range of the many arts of living. Like the complex political, technical, and economic dimensions of current modernity and progress, the city requires its own theoretical validity, its own inner harmonies, and steady advancement on numerous fronts.

Such a position on cities will strike many readers as errant and grossly misplaced. To put cities at the center of society's threshold of a new humanism does not coincide with any existing clear momentum in society, one might say, or with what most people consider important, let alone central. My answer is that perception itself may be the issue, for progress today in the social arena is, most of all, a matter of concept, not invention. Action can follow. Modern industrial society has achieved, if anything, a paramount ability to act. Simply put, society can do pretty much what it decides to do. Therefore, our horizon and the critical question in our day is our perception of the greater possibilities of life, and these must revolve around organizing a new range of freedom. Again, that means the city. Yet the lack of importance given to cities today speaks for an urgency on their behalf, to create a new foundation for humanity while also resolving many major issues of society.

Cities fare badly under the dominance of economics in development. In many respects, they have retrogressed under the impact of machines because their power of organizing development has not been recognized, especially to promote the higher human potential.

Society badly needs a new inspiration to supplant the overly simple onward and upward struggle of raw output now giving us all so much fatigue and frustration—and the consternations we have in confronting so many social crises.

Misdevelopment

> *Modern technology...has failed us...mainly because it has forgotten its basic function, namely that all technics are, in the last analysis, the tactics for living.*
>
> Henryk Skolimowski

Now we see a new oppression bearing down upon us. No despot is evident. But there is an invasive and pervasive tyranny built into the greatest depths of organized life. This is both ironic and critical, because it is buried in everyday life, interpenetrates our motivations. And even boastfully claims a close affinity to what life is all about, calling forth affluence in defense, even its apparent freedom, and the massive success of free enterprise.

But the reality of our modern oppression is readily seen in the myriad crises across the land, headlined in newspapers, reviewed in academic and professional journals, and bewildering to everyone. Youth seems to become a synonym for rebellion, drugs, sex, and crime against the sound of a rock band and is hardly any longer a mere coming of age with a naive indulgence. The culture of crime advances, spurring the greatest expansion of police and boom of prison construction in history. We try to solve traffic congestion by promoting more traffic. The stock market reaches new highs while corporations downsize, poverty and homelessness grow, a new class of working poor arises, and healthcare costs have often become more immediately cause for trauma than death. Therapy grows upon job insecurity, concerns for bodily safety, marriage, a flotilla of interpersonal anxieties, and an indecisiveness about what life is all about. On the one side we furiously race for affluence while the real benefits of affluence recede almost in proportion to the growing GDP. Still, while inflated abundance is real, it echoes an emptiness. On the other side are the sores, boils, and scabs of the society, which are but the superficial symptoms of a systemic social disharmony.

Seen in the long perspective, the racing momentum of industrial progress now follows a course of increasing contradiction, of forced economic development bearing less and less upon human need or independent personal desire. In its place, the economy grows upon new forced necessities and the growing social debilitations of those necessities.

The unrecognized side of our modern tragedy is that an increasing part of the social crises we face is directly a result of economic progress. Here is the overwhelming fact of modern misdevelopment. Here is the pervasive fact of the new social oppression. Here is the deeper loss of freedom masking as the growth of freedom.

Perhaps until World War II, we could justify saying that any given new kind of problem appearing in society with new productivity was the price of progress itself. That is no longer a reasonable response when increasingly acute issues avalanche to overwhelm and even disorganize society's ameliorative processes. Economic development, in other words, is on a collision course with many of the better public interests of society. Traditional industrial progress continues, but much of it steadily becomes literally counterproductive in deeper human terms.

The collision center of traditional progress and the consternations of progress are, of course, in the cities. They encompass the pockets of the best and the worst of our modernity, where strength invokes weakness, where growth and decline all fuse randomly or chaotically together. But mostly, they are the battle zones where sacred, protected competition and profit prevail over non-sacred public betterment. Without new guidance, cities are becoming a self-inflicting ecology of disaster. Without a growing and effective leadership or movement, at least equal to the conservation movement, cities cannot achieve either a constructive humane or natural order. They then become the disaster for both human and natural ecology.

Cities today are becoming a created congestion arising paradoxically from massive sprawled dispersion. This multiplies the necessities of urban travel. Together dispersion and congestion promote both urban decay and further motives to escape to the distant suburbs or beyond. Then dispersion is escalated, and the congestion of traffic from dispersion grows again. Tragically, the economy thrives upon this spiraling process. Clearly the great efficiency of industry promotes the greater inefficiency of cities.

In human terms, affluence is becoming another name for misdevelopment. Increasingly, big money has become a cover-up for material waste, social disharmony, and promotion of a wide span of public and private crises. Waste, disharmony, and crises are prime conditions underwriting more economic opportunity—ironically, more "wealth" for a growing GDP. Thus we create a corrosive inhumanity in the name of product and service, an oppression that binds us to an ever-increasing consumptiveness.

It is especially critical in this time of misdevelopment to beware that the chief faults of a society closely parallel and often arise from the chief strengths of that society. In the United States, we see how the private sector creates many conditions of crisis, but then politically prevents public solutions, aside from those that promote further economic growth, such as highways and airports. Contrary to prevalent assumptions, it is nearly impossible for the marketplace to establish integrated urban patterns. Capital must seek profit, and inherently it cannot solve the broad and combined problems collectively created by the entire marketplace. Here, I believe, is a central contradiction of our modernity.

Ironically the end goal of economics must ultimately be less economics. This is as true as the best reward for efficient labor is less labor. Yet as basic as this may be, such a prospect has no broad public or economic role in today's society. The endless promotion of excess product remains the order of the day, for we create new necessities that force us to increase consumption.

The answer to these and to so many vexing questions of our times is a new kind of city, creating new social efficiencies while responding to the rainbow of human aspirations. New cities are necessary to integrate social functions for full human benefit and to relieve the current domination of economics. The economic dominance of life will one day inevitably give way. It has been only two centuries in the making, and its power can dissipate in much less time. The important question is how creative the new foundations can be and how soon they can take on a more positive social role. For this reason, the city will likely become the most significant arena in the coming period of positive human growth.

We have yet to build truly democratic cities. When we do, their structure, function, and social organization will open grand social vistas for personal development.

Expanding Behavioral Freedom

> *A new polarity emerges: a day-by-day insight into the tension between the manipulation of things and the relationships to persons.*
>
> Ivan Illich

The great promise of the future city is to create a rich physical and institutional setting for the expansion and diversification of the human experience. That city will offer the immense repertoire of civilization to everyone to build a much larger, more meaningful, personal experience throughout life.

To expand behavioral freedom, the urban structure requires the purposeful development of three spirited arenas of living: family, community, and cosmopolitan life. Together, they can optimize experiential diversity.

The modern spirit—virtually the invention of modernity—is a cosmopolitan engagement of persons with the whole society: instant communication and jet travel, finance and technology, bureaucratic organization, hierarchical authority, personal qualifications, and precisely measured success. Involvement is calculating and cold for those who risk, usually perfunctory for those who follow in the stable niches of career.

In the modern world, all that is not organized into life is in danger of being organized out of life. Hence only fragments of community continue to exist in cities. The family is under siege. In public affairs, we are dangerously close to operating on the assumption that the professional cosmopolitan life—a life predominantly economic—is the only ball game that counts in life.

In other words, we have created a great cosmopolitan experience over the last two centuries and left behind the more precisely human realms of family and community. Both rapidly declined when they lost their functional values of organizing production and trade to corporations in mass urban society. As diverse as the cosmopolitan world may be, it operates strictly on functional performance and monetary terms, and hardly promotes the deeper personal values we cherish in life. Involvement is organizationally impersonal and conditional, for all persons are expendable to the world of cosmopolitan bureaucracy. The life cycle of persons is subordinated to their careers. Only the wealthy and the pensioned can confidently avoid the insecurities of the cosmopolitan life.

But in the future city, the cosmopolitan life can be structured to be part of a greater social scope of human organization. The city can also provide a secure place for stable and healthy family living. Here the intimate and emotional foundations of personal security may be combined with personality formation and educational roles in raising the young.

The support for these roles is logically found in the immediate community, a crossroads between family and cosmopolitan life. A modern community can offer persons a range of experience far exceeding anything associated with traditional community. The promise of a modern community is to rebuild the associational heart of public life and create a new interpersonal realm for highly varied social participation. The spirit of community can be a casual, festive, and a creative counterpoint to the cosmopolitan world. It is the realm for the amateur, one who pursues an activity for the love of it, and for a creative social serendipity, a vitality of constant surprise and revelation.

Community is local, based on residence, and encompasses as many services and activities as are manageable locally. Community is highly participative, self-managing, and unified, offering a finite local identity for the person and a basis for common action. The community council becomes the corporate instrument of the people, serving their expressed personal and social interests. It can re-create direct democracy, bringing public affairs down to eye level.

Whereas urban industrial civilization has created a new performance-dependent personality attuned to the cosmopolitan world, a resurgence of community may build a rich and diverse foundation for a wider range of personality, something like old timers called character. The future city that expands experiential diversity and wide opportunity for people will therefore stand on three broad foundations: (a) the infinite functions of the cosmopolitan world, (b) the social strength of community, and (c) the intimate relationships of family.

Every person may define the manner and extent of engagement within each of these realms. Some may associate exclusively in one, some in two, others in their own way with all three. Each of the experiential levels has its kinds of human performances, attachments, rewards, and emotional or intellectual responses. Each can serve persons in different ways at different stages of life.

The three spheres will let us structure our cities and organize our lives around the best of our humanity, the richest excitements of our diversities, and the magnificence of our creativity. These things can be done together in cities, collectively, institutionally, environmentally, and with the grandest spirit of every one of us.

Chapter II

Society's Master
Mechanism

It is in and through the city, with all the resources it offers for the mind that man has created a symbolic counterpart to nature's creativity, variety, and exuberance.

Lewis Mumford

Idea of Community

The clearer and more defined a group becomes to a person, the easier it is for him to concentrate his commitment there.

Rosabeth Moss Kanter

In supposing to create community in the modern metropolitan setting, it is important to be clear about the term community, some of its history, and what might be expected of it. What was community in the past? What is its general meaning today? And why should we attempt to recreate it—and create it very differently in today's setting? Are we being merely sentimental about a condition of life that, at best, may have existed only partially or only in very special conditions, such as a small hunting society building harmony for coordinated hunt or perhaps a farming community harmonizing its practices to irrigate land?

I define community today as a local, residential, interpersonal, self-reliant, self-determining, and integrative human institution promoting a full creativity for expansive personal experience. While the essence of human growth inevitably derives from interpersonal behaviors, the person must always have the right of entering into community essentially on her or his own terms and exiting community at will. Regardless of the history and meaning of community, one need not pry deeply into philosophy and psychology to know that the entire basis to build a good society must be centered on the essential idea that people derive their ability to function, their personal stability, and their humanity from the mutuality of direct interpersonal association.

We have created numerous public institutions to serve specific human purposes (e.g. schools, hospitals, prisons) and voluntary associations (e.g. Elks, Rotary, Lions). We have yet to create a modern institutional framework that serves the whole person and whole social group in a broad range of activities and services that are integrated around the person.

In some respects bureaucracy itself is a modern extension of traditional community. Both community and bureaucracy do effectively and rationally organize work. However, corporations ridded themselves of the common well being of individuals and substituted simple wages in its place. When the deeper human meaning was eliminated in favor of proprietary interest, the term *community* could no longer apply to an organization confined to working for wages.

Traditional community was interpersonal, identified with an ethical behavior, functioned through a traditional rather than a strictly rational form of organizing behavior, and was usually associated with kinship and clan. That kind of ethical behavior, however, in bureaucracy has now become associated with nepotism and therefore corruption. It thus remains a source of conflict in almost every modernizing society and is illegal in government. As a result, the traditional values of community and modern values of bureaucracy run deeply against each other.

All of bureaucracy is corporate, either governmental, business, or nonprofit. Community itself has not been incorporated, that is, not established in the hard-hitting, self-defending manner of modern organizations. One might call community a soft institution, since it does not have the wherewithal, the centralized direction of action, or the focus of leadership, to defend and perpetrate it's being. Classic community, in other words, is clearly not an institution that would land in court or start a war.

Perhaps the common, casual, and endearing qualities of the traditional community are like the high qualities of personal character that emerge when that person is thrust into trench warfare. War demands warlike behavior, regardless of individual inclination. Sterling qualities of persons or communities are simply overwhelmed by the forces and necessities of the time. What matter is it that a person possesses great charm during battle or that a community reveals human comfort and security during a startling explosion of materialism by industry? Thus the effects of intense practicality and aggressive personal ambition (fed by productive wealth, hierarchic power, and anonymous urbanization) also undermined the soft, personally endearing interpersonal relationships of the old community.

When physical labor was arduous, the hard work was often accompanied by common song in the fields and in beating time while hoisting sail or pulling oars. Now that hard labor is rare, the common song is gone, and music comes from outside of us, entertains us, but is hardly a part of us. And too, it is, I believe, with the greater richness of human community: And so, if such a community be a myth, as in some degree it certainly is in our varied human past, it is nevertheless a yearning myth upon which we can create a fine new reality.

I believe that a soul in search of community resides in nearly every person. Since we are just as social as we are biological, that craving of our soul resides more longingly within us today as we all face up to the great and little treacheries of urbanism and bureaucracy, to the turmoil of getting through each day, or in combat, on our wits, all alone, at least alone in spirit.

That soul resides today in the continuing wide use of the term *community*. Unfortunately, it also was applied very powerfully as *commune* and *communist*, which resulted in many tragic decades of horror. But now history suggests new beginnings, one part on the ashes of communism, the other as redirection of the forces of capitalism. These bear down upon us with a peculiar suppression that we have accepted all too willingly as the price for our rising productive wealth and our exotic materialism.

Our aim, therefore, is to restore community, give it day-to-day validity in life and social idealism, and incorporate it in the modern fashion to be at least on a par with other defensible organizations. But more, our aim is to give community new purposes and new functions that are particularly modern and to put human beings squarely on the forefront of creative social change.

Community, I believe, is especially suited to translate change of the twentieth century into a flourishing human greatness never seen before in any civilization. We can transform one of the oldest informal institutions in society into one of the most modern, formal, and socially idealistic organizations of human life. By giving community new active ingredients and its own institutional self-reliance, the roles of community may then revitalize social life profoundly.

Like almost everything modern, valid, and useful, community must be consciously conceived, purposely created, and carefully organized. Most human beings can easily observe that both the society and the character of persons are increasingly self-made, that is, no longer merely customary, slow changing, and personally rigid. Whatever the nature of the soft human clay we are born with, today we shape and reshape that clay of and for ourselves and others from generation to generation. Currently, we shape ourselves to be *productive citizens*, to produce and consume profusely within the bureaucratic mold. Now perhaps we can aspire to become *creative citizens*. And upon such an image, community can be visualized.

Community offers a new way to shape strength of character. One of the outstanding facts of contemporary development over the last century is that in remaking society to be productive, we have been refashioning the core of human personality also to be productive (and now consumptive). Human nature may not change, but how persons respond to their vital social imperatives shapes their personalities. Community offers us a very diverse way in which the social side of society may shape itself consciously and positively to a model of our own liking.

If the human result of community is a freer personality, that person will be more productive as well, especially in work demanding broad knowledge and critical judgment, if not creativity and initiative. Adjusted, self-confident, and spirited persons will always contribute most to society.

Modern community thus takes the historic sense of community, integrates and unifies the bewildering complexities of urban life completely for the person, and becomes an institution for the positive evolution of persons and social life in a process of self-development.

We might think of community as a kind of corporation created specifically for people *as persons*, assisting them in their self-development, not to produce and market specific products or services. Thus community shapes a wholeness, strength, and varied service for the individual. It simultaneously forms a sophisticated social *taproot* out of the *grass roots* we now refer to in the fragmented character of social action. And in the modern world of highly organized blocks of special-interest power, community offers people a corporate power to stand up to the field of powerful organizations. While community gives the person new worldly strength, there is also created a high-potential social, cultural, and natural setting within which the person may set out on a course of human magnificence.

Community speaks to the wellspring of what are most human about us, not for everyone, not for anyone all the time, but for most people in hundreds of ways. Unlike the past, people today can choose, choose community—or not—or move from community to community as each unique community's character might suit one's personality or stage of life.

Can a community become dull and overly routine? Indeed it can be dull if it is conceived fractionally, considered as merely a different setting for one's dwelling, or entered into by its residents with the same attitudes that exist in filling housing needs in cities today. A dominating emphasis on costs and class structure will be damaging. For a community to succeed, it requires some common aspiration by those involved. Every person needs to seek and to give and to think of personal self-development and community self-development together, as certainly as each strengthens the other. The vitality of the community is a learning experience, perhaps like a marriage. An expectation, a sense of experimentation, a reasonable commitment, and a willingness to trust anew—all of these will improve the likelihood of success, no less than other kinds of human enterprise.

We talk of genetic engineering and hundreds of other frontline scientific advances. But rarely do we hear a whisper of focusing the same level of resources and dedication in building vigorous, creative, and comforting social organizations for the broadest range of human benefit, thus building a new vision of democracy along the way. But today we still adhere to the dogma of arduous productivity that has dominated thought for the last two centuries.

Community may be considered an empire of the person, a physical and social environment that articulately serves a wide compass of human desire—at a scale, composition, and character for the person—a self-empowering union of persons, a guild of the amateur. Most people need to belong, to share and cooperate, to personally give and receive of and for themselves, to be honored and appreciated for large and little things among friends and family. Community can be an instrument for such an empire of person to grow.

Inspired Whole Cities

The words city, civic, citizen, and civilization all have the same root and identify historic human progress with the city's own development. It has never been otherwise. And while being the foundation for civilization, the city itself also is the greatest of civilized artifacts. Cities are quite justifiably the heart, mind, and inspiration that make civilization what it is.

But despite the obvious power of cities to propel civilization, the cities of the modern world suffer a severe melancholia of spirit, while being treated as merely the warehouse of development. We do not find the great visions of human environments with integral facilities to work, play, educate, care for health, or serve the elderly or handicapped that we so diligently integrate for space programs, military activities, and industrial output.

The profound depravity of the urban spirit essentially reduces itself to a depravity of people, which exposes a grave lapse of democracy. Progress is more than affluence, and in the long term, something very different. Progress also must expand human opportunities and freedoms in dynamic environments that will be built with at least as much system of thought as we now put into building a great airliner. Hence we must look at cities anew if we are to envision progress in a world that is now becoming safe for democracy. A new civilized inspiration for cities will reach farther and deeper into the personal psyche, social dynamics, art, and science in society. True human progress lies there, built on the foundations of technology and economics, to be sure, but rising atmospheres above our present conception of life in cities.

I dream of the city, for I find no other vessel of society capable of fulfilling the human dream of a better life. I can dream only because I am social and because the social ingredients of my dreams are made in the city. But dreaming about cities is not like dreaming about fame, fortune, or love. Cities are complex human environments that can creatively unite—and therefore magnify—the special benefits of a thousand otherwise separate urban functions.

But tragically, they now reduce or destroy a large part of that creative potential. Society fails most in finding creativity of the whole city. The reasons are many: the rural character of individuality carried forward from our past; corporate pressures for an unobstructed urban playing field to speculate upon; a belief that supply and demand (a higher standard of living) alone can fulfill human need; a defensive privacy-and-escape mentality generated by uninspired and undesirable cities; and the near absence of a visionary tradition of citymaking. Modern cities themselves, unfortunately, are seen largely as problems, and our outlook upon them is tragically negative.

In such an atmosphere, it is not surprising that problems have become overwhelming, including public services and taxation, conflict of movement and the peace of neighborhoods, slums and drugs, crime, malediction, and ill health. Those who can afford it escape to safe suburbs, demand tax moneys for freeways, and in the process create a wider gulf between the haves and the have-nots, the city and anti-city suburbs.

Thus the very places we build to spend our lives in have become the polluted backwaters of society, the places instigating escape, first to the suburbs and open country, and second, to the mountain or shore. The cities we build are powerful machines producing magnificent products and services, but do so in environments of wasteful debris, a degenerative life setting with growing human conflict, and for so many, simply hopelessness.

Where is the good city to be found in the ideals and theories of society? Not in economic development, for capital investment speaks inherently and exclusively to the generation of dollars. We have a social system today that operates on two levels. The first promotes economic growth for profits and jobs. Our Council of Economic Advisors speaks positively for government to the wholeness of national economic conditions. The second reacts to the many resulting problems that force themselves upon society. They speak painfully to each crisis, one by one. So we see how the cities of the 1980s 1990s foundered in crises while the economy grew handsomely under supply-side economics. So cities that struggle with problems and promote escape to the suburbs up to a citizen's ability to pay can never be cities of inspiration.

Sadly, the founders of America never got around to the creation of cities. They can be excused, since their largest city, Philadelphia, had fewer than forty thousand souls, today about the same number of people killed by the automobile each year. Yet more recent leadership has had little more to say about cities, even as those cities have become the crushing presence of modern delusion. Speaking to the social elements of the city, politicians address separate issues of working conditions, child labor, sanitation, housing, civic monuments, fire and police protection, freeways, and airports. Sometimes they have bowed to general planning, as long as they could later ignore it. Subdivision and zoning ordinances were adopted. These helped, but they isolated functions, bloated transportation, and consequently fostered a larger array of urban disintegration. The lack of a unifying urban methodology has led to possibly the greatest array of crises in modern domestic history. Our attention nevertheless remains focused on the isolated and negative urban issues, not on the whole urban potential.

Philosophers have served society probably better than most people are aware. Yet they too have had little to say about the enormous urban dimensions of our industrial civilization. Our modernism will always have a gargantuan void until there is founded a vibrant, positive social élan underpinning the person and all that makes up the city. The only real heritage of our great and growing technology dwells dormant in our bodies, minds, and lifelong aspirations. The challenge of our time and the promise of the wholeness of cities is the creation of a grand spirit of urban life.

While growing cities have served as foundations for industrialization, our great opportunity now is to transform the industrialized urban structure into a rich social nutrient for every person. With advanced technical support, massive new opportunities can energize the inner and outer greatness of our being, giving each of us new freedom as we individually make it.

Cities have always been the master mechanism advancing the objectives of society and spearheading the development of civilization. We can trust, I am certain, that cities can now become the master environment advancing the social dynamics of society as far as it has propelled the industrialization of society.

Redefining Progress

> *Society cannot be reformed by creating more wealth and power. If it is true…that our environment and way of life profoundly affect our attitudes and those of following generations, nothing could be more distressing for our immediate and*

> *distant future than the decadence and ugliness of*
> *our great urban areas.*
>
> Rene Dubos

This age, inventive beyond all others, also exhibits huge contradictions founded on that inventive force. That is ironic because the chief strengths of our age also are rapidly becoming the chief ills of our age, resting, as both do, on rapid change and society's concentration on industrial output and commercial promotion.

The incredible trail from science to technology to product was achievable only because it also was supported by a bureaucracy, massive capital markets, and commercial distributive networks. The critical juncture was between technology and bureaucracy, and this was the work of corporations, which organized an unprecedented range of resources for maximum return on investment. Not surprisingly, today the great glass towers (bureaucracy) and huge, low-lying plants (technology) chiefly symbolize the dynamics of our age, and the concentration of modern wealth and power.

The linkage of our age to money and economics is also like no other. We cannot, it seems, have a social action without a financial transaction. Great flows of money propel great flows of resources, labor, land, product, and service. Our dominant system in society, economics, demands that all things manageable be organized by this common coin.

We have elevated the process of economics so far that we promote only economic growth and do not seek to understand the full range of economic consequences. All things encompassed by economics are raised to the highest level of social legitimacy, virtually without concern for their social effects. Public affairs have functioned on the principle that market economics is the master mechanism of society. Certainly, this claim had great merit while the system was building the earlier foundations for great productive wealth. However, carried to our time when the wealth is causing broad consternations, the claim precipitously loses validity.

A true master mechanism must do more than balance supply and demand. It also must give a sense of direction to change, and that requires a broad goal-making capacity, upon which the market falters. Leaving individuals to choose in the market the things that determine their course of life very easily damages the non-economic values in life. Without new categories of choice being created, especially in cities where non-economic organizations were not able to shape a sustainable social life, economic choices merely divert or actually defeat the greater social possibilities.

In reality, the city should be the master mechanism of society. That case rests on a number of factors: (a) the city is essential to all industrial and commercial activities and determines many of their efficiencies. (b) The city determines what products and services are required and appropriate in society. (c) The city organizes the essential human ecology on earth. And (d) the city is itself the environment that establishes public and personal efficiencies, interpersonal and social opportunities, educational and recreational facilities, the aesthetic qualities of built and landscape spaces, and the places of tradition and festivities.

These are the choices that can only be made in common, that is, made socially for social purposes. As purposes, they lie outside of the scope of market economics. And they are now choices denied in cities organized for only market choice and private wealth.

Only cities shaped democratically for broad opportunity can be realistically considered as the master mechanism of society. But first, that purpose must be understood and pursued.

Like all institutions in times of great change, economics will inevitably transform or be transformed. Pressures for a better kind of performance run deep. Such times suggest both the necessity to basically reconsider what constitutes social progress, and the opportunity of stimulating a new social awakening—comparable to the transformation of outlook stimulated by the industrial revolution itself. These times inevitably reach to the fundamentals of democracy in an urban, technical, and corporate society, especially now that imperial communism no longer complicates the issues.

A system flushed with success may antiquate itself as thoroughly and as rapidly as failure may defeat it. Successes easily lead to excesses, as they do so disastrously today in cities. Or successes may offer society a basis for a higher stage of social development, as they also do in cities. Either way, fundamental change is inevitable.

The classic definition of progress, centering on a high standard of living, has been the dominating national ambition. To a large extent, that ambition has reached fruition. But that ambition, with all its success, has run headlong into many useless redundancies and hard-rock contradictions—including a significant part of society that does not share the largesse of great productivity. If new goals are to bring us to a new definition of progress, that definition will go beyond dollars, beyond product and service, to permanent achievements of built environments, preservation of the natural environments, and the formulation of vast new opportunities for people to learn and associate. Cities built in pursuit of a new progress certainly deserve and invite the same range of creativity that society has so singularly given to technical invention, enterprise, and product.

Urban Focus of Life

The promise of cities, like the historic promise of science and industry, is comprehensive to the human condition: for all persons, for a valid human ecology, and for simply greater effectiveness and efficiency of persons in everyday affairs. And once cities begin to substantially liberalize and reform technical processes, we may expand confidently toward the infinite variety of human experiences and excitements.

The largest and perhaps most central opportunity will arise when an integrated urban environment brings people together to rub shoulders, so to speak, in ways in which hundreds of social, recreational, cultural, and other activities are conceived, organized, and undertaken with great ease and enthusiasm by people of like interests. Conversely, today's cities are organized to render exponential hardships for people pursuing their common social interests. One only need mention long distances, travel expenses, time, and the lack of common and congenial meeting places to understand the severe associational deprivations now present.

New cities can literally underwrite a new social nature through urban design, building an ease of interaction, efficiency for cooperative endeavor, and simply an easier way for one to get to know other persons. We can create viable urban communities. If we bring creativity to the organization of community, a thousand kinds of human opportunity are invited. We will view modern community as a multipurpose organization that is comparable to the modern corporation. What we have done for corporations and their purposes we can do for people and their purposes—with all of their diverse interests. But we require vision and will.

Beginning Anew

Let there be no mistake. Cities are not merely sites to propel economics. Nor is the optimum city to be made of better-looking streets, social facilities for old folks, or a suppression of crime, that is, projects that make the city merely livable.

We aim for, and we can achieve, fundamental and very positive changes in the anatomy and physiology of cities that reflect ever-expanding values of life. A new wellspring of human inspiration can be foreseen, one that is wider and deeper than anything we now know. But we cannot predict specific new behaviors or aspirations of life any better than, even as late as 1900, we could have predicted jets, computers, and synthetics.

Although the specific content of a new era is for future generations to experiment upon and determine, a more comprehensive social freedom can stimulate a new human flourishing and broader social vision. Some possibilities are easy to see: preserving traditional cultures, creating many mixes of art and technology, building lifelong learning, initiating many individual and small-group projects, and establishing new forms of association and interpersonal living.

Certainly, the foundations and expressions of a new human flourishing are necessarily close to the individual—where we as finite human beings eat, sleep, and may either be imprisoned into sterile lethargy or freed for many opportunities. The many benefits of both the built and the natural environments can be greatly expanded close to everyone through an optimal urban geometry. Finally, these possibilities can be achieved while also lifting the heavy overburdens of today's inefficient and socially barren cities from every person and avoiding the epidemic of urban crises.

Reformed cities, when they become a central objective of society, can positively broaden the way the public thinks and acts, even expanding the way individuals look upon their own lives. Inspiring urban goals seeking out the widest reaches of the good life will replace the singular pursuit of economic growth while narrowing the reactions society currently makes in facing the myriad urban crises. But today, note how society views the two great arenas of public affairs. While we set economic growth as the central, positive, expansive, and endless national goal, we conversely look upon cities as problems and try to push them aside with narrow, specialized palliatives. One we seek out in very positive terms. The other we react to in very negative terms.

Consequently, many of the most costly programs, ostensibly for the public, in reality largely serve specific sectors of the economy. The interstate highway system (now increasingly intra-city) was built principally as a boon for numerous industries: auto-making, fuel, trucking, construction, motels, restaurants. While there are also sizable benefits for the public, one doubts that Congress would have passed the system for the public benefits alone. The industries saw great expansion and profits, and they vigorously promoted it in the early 1950s.

Compare the interstate highway development with the then concurrent program of urban redevelopment. This latter program was limited to physical conditions, confined itself to the removal of blight, generally excluded social conditions, and was funded at a level to assure failure. The conception was mostly a specialized response to a specific problem of housing. No urban ideal was expressed, only a problem, the elimination of blight. Even an overtone of welfare—giving something free to the poor—was enough for Congress to narrow the concept of redevelopment. In the end, a powerful and far more costly program of freeways could enter into cities, divide neighborhoods, and promote blight with ever more traffic disruption, air pollution, and noise. In one period, the same number of urban dwellings were moved or dismantled as an incidental result of urban freeways as by the entire program of redevelopment.

One might properly label the stark differences between *positive* economic goals and the *negative* social and urban problems as a case of social schizophrenia. Certainly, this immense torque on government, corporate, and private affairs is embedded deeply in the mass media, the range of education, the character of the public mind, the nature of lobby politics, and the direction of private and public investments.

But, as certainly today, the raw urban and social problems ominously creep up to challenge the entire range of economic progress. This is where the positive and negative sides of our national schizophrenia bite deeply into what we all assumed was the *social* worthiness of free enterprise.

In emphasizing ideals and goals of cities, society will simultaneously get close to the root sources of crime, family disintegration, public health, transportation, environment, and other pressing matters without losing sight of the immense, positive human benefits that are possible in cities. Most simply, the issue of good cities reduces itself to whether we chase—and expand—crises, or whether we pursue a broad range of positive social goals, and by this means resolve many immense social issues.

Cities can be built for enormous practical benefit. We can preserve as much as 80 to 90 percent of urban land for a very wide range of recreational and environmental benefits. We can then conserve as much energy per capita, and this will allow a larger part of the remaining energy demand to be met by renewable sources. Similarly, all utilities will have shorter lines, be more simply organized, and be less costly.

The future urban environment can reduce most commuting to a casual-ten-to-fifteen minutes while virtually eliminating hassles and costs of cars, freeways, boulevards, and parking facilities. Traffic danger to children can be all but eliminated. Pavements, harsh symbols of environmental brutality, will be radically reduced. The future natural environment in cities will be spacious, varied, easily accessible, beautiful, and even highly sociable. Based on local desires, it may include natural forest, play forest, and *urban farms*, as well as golf courses, formal gardens, playgrounds, and varied parks, all incorporated in one unified close-in design. Individuals may garden as much or as little as they wish and change their wishes at will. There will be little need for fences, the dominating icon of defensive privacy. The high proportion of trees and greenery in conjunction with greatly reduced energy use will help clear the air and reduce ozone depletion. The rich variety of outdoor environments will provide equally varied opportunities for on-site and in-depth education on seasons, soils, biology, and ecology. Central and simple recycling and composting will virtually eliminate garbage dumps.

These urban possibilities are not pie-in-the-sky predictions. They are practical and achievable in about the same way that the standard of living grew, except that in cities the efficiencies and qualities are measurable and predictable in advance. The major burdens will be the costs of development. But the critical element is a *public vision and will* comparable to that which created today's material wealth.

We are talking about urban environments of enormous efficiency. Like factory efficiency, urban efficiency will be founded on layout, which, like manufacturing, benefits enormously by reducing the need for movement. Today's cities are now disasters of movement—disasters because the most inefficient, consumptive, polluting, dangerous, and costly form of movement, automobiles, have been made into absolute necessities. Industrial societies have not yet begun to confront the monumental issue of destructive *automobility*. The issue cannot be faced without an attractive urban alternative, of course. This can be done for everyone's benefit. As we shall see, the choice is not the prohibition of automobiles in cities. Rather it is to be found in *eliminating the necessity* for automobiles in the city, indeed in greatly reducing the need for formal transportation of any kind.

Upon one of the greatest complementarities of modern times, we will find that the highly efficient city also is very kind to the physical sensibilities of the body, delightful to the eyes and ears, and most of all, inviting to richly varied sociability. With an urban efficiency that can fully match industrial efficiency, the city can break the meaningless, but arduous and wasteful, production-consumption syndrome that is at the base of so many modern problems.

Social and Ethical Frontier

How cities are made tells us how we are made.

The city's power to make people what they are emanates from every part of the physical environment; the organization of industry, commerce, and transportation; the design and facilities for social, religious, and cultural pursuits; and varied associations. And it emanates from the familiar beauty of local environments that create fond memories, stir emotions, give a security of belonging, and offer a comfortable sense of past, present, and future. The arts, values, habits, associations, and traditions associated with favorite places and people can be created together to underwrite a broad, new social ethic. And so it is that people are beautiful in beautiful environments, happy in home settings they can manage, excited in places of diverse cultural participation, and secure where all essentials of life and society are immediately accessible.

Conversely, when people are organized sharply toward economic rewards and threats, their values become weighty with dollars and cents, imprinted upon them and into them. All too oppressive is the linear value of monetary success or failure. Strangely, we confine ourselves to and struggle almost singularly for success when there are endless other values more worthy of our efforts.

We can design cities for very broad and open-ended social values, just as our constitution designed the essence of our open-ended political values. We can develop an urban constitution that will shift emphasis from making the individual merely a productive citizen to creating a broader, fuller, and far more creative person. When we do make such a fundamental shift, we will find, I am certain, not only a person of far wider outlook. That person also will be simply more productive.

Social equality of opportunity can benefit as well. While we should always deny an elitism of privilege, we should always promote an elitism of performance, that is, a personal and social élan. Perhaps we cannot generate a great new Leonardo da Vinci. We can certainly generate many little Leonardos.

The purpose of new cities—and possibly the most central purpose of society—is to unleash the human spirit and to prove that human nature is what humanity makes of it. Perhaps our self-created nature can be the greatest of all human creations.

The collapse of communism lifted the iron curtain of systematic tyranny for many millions of people. Still to be dissipated is the Marxian pall that has hung like thick smoke over the social thought of the Western intellect, especially that of social scientists, who out of fear of a new McCarthyism have rid themselves of relevance or creativity in so much of their work. Such a pall has never afflicted physical scientists and engineers who do their experiments precisely for application in society. Now it is time for all social thinkers to do the same and to break the logjam of thought about human beings in a world of revolutionary technologies and rapid transformation.

Cities are central to everything modern, even to the idea of wilderness. In personal terms, cities form the commonwealth of society. Through its actions in cities, society will proclaim that it cares for people in a far fuller scope of democracy. In this age of new beginnings, new challenges and opportunities are emerging. The critical question is whether the best of collective social intelligence and the strength of popular will can build cities that are unconditionally good for people.

Building Good Cities

I speak of the city, which I now mourn and in which I find the central locus of human promise. Yet there is tragically little awareness of both the grave condition and the great promise of cities.

The city is comprehensive to the well-being of Americans: physically, technically, economically, ecologically, politically, and socially. Since the challenge is comprehensive, so must be our thinking and action. When the broad, positive powers of the city do become widely known, we will appreciate how deeply they can nurture or deaden emotions, excite or suppress intellect, protect or threaten personality, and determine even how our eyes may meet. These great urban powers now operate incidentally, or wildly, or dangerously. They are not yet matched by creative purposes and imaginative frameworks to assure outcomes that will solidly benefit people. Constructively organized, cities can form the basis for a broad, new human initiative in society and a vast, new scope of democracy. Cities may best help us become creative about past creative inventions and help guide their cumulative impacts in society, avoiding large and small catastrophes.

Whatever future technical progress may occur, we can proclaim that the future is social and fully human, not to be overwhelmed by technology or bureaucracy. To do so, we should tack our ninety theses on every door of science, enterprise, and government. All society can learn that, indeed, the city has become the chief question of this age. Everyone needs to know its boundless possibilities.

Life is as exalting as we make it or as base and bestial as we make it. In the complexity of life in this age there are many dead ends and many possibilities for bestialities to evolve. These may even appear normal and right, sometimes in the name of progress and freedom. Society can create antennas that will signal the many early forms of blockages to, or diversions from the public interest. To do so, we all need to be very clear about our values and goals, for they offer the best means to be definitive and to sort out bogus, diversionary, excessive, and disruptive developments.

The essence of this book is that we are now in the first historic circumstance where a substantial part of our creative and productive wealth has become seriously detrimental to human well-being. The irony and consternation is one of progress itself. At the same time, other creative human endeavors have become both necessary for relief and urgent in defining a new direction for progress.

CHAPTER III

ORGANIZED DESTRUCTION OF CITIES

Nature is vanishing. The city is vanishing. The accelerating dissolution...has induced a massive compromise.... But both the pseudo city and pseudo country, with commuters shuttling between them in a desperate search for satisfaction, which neither can provide, appear in the end to promote little more than discontent.

Serge Chermayeff
Christopher Alexander

It was as revealing as it was disheartening to walk out of orderly, efficient, beautifully managed factories into a turmoil so shocking...Here was the visible proof of the gap that exists in our society between our mechanical genius and our social ineptitude.

Agnes Meyer

Why Cities Fail

Cities, I repeatedly stress, are essential for all higher human possibilities. They must be recognized for their utter supremacy in defining modern life. Recognizing them for anything less decries democracy, defeats industrial progress, and leaves us dumbfounded in the whirlwinds of the uncontrolled change we ceaselessly unleash. Cities define the nurturing conditions, qualities, ambitions, intellect, and verve of the young; the challenges, opportunities, and momentum of the mature; and the fulfillment and peace of the elderly. But the cities we make are defeated in their very creation.

I do not exaggerate: *Defeat of the city is defeat of society.* We dare not belittle the instrument so pervasively essential to all civilized life now so vanquished.

Whereas change of the last two hundred years has been transformed increasingly into economic opportunity and previously unimagined economic power, a different kind of change coming into view is based specifically on the human qualities of change, not the mere quantification of product. Given new directions for imagination, a new public aspiration and social commitment, cities can create another previously unimagined greatness, possibly as far ahead of our dreams today as modern technology was two hundred years ago.

Since times of serfdom material progress has been absolutely astounding. But as much as we benefit from the industrial miracle, the economic power has been shaped in recent decades into an invisible social prison, a complex confinement showing the many attributes of exploitation in the vast wilderness we still call cities. But with new vision we can look upon the great economic structure now in place as the foundation for another more human vision of society. The more readily we recognize the economic confinements and demeaning change, the more readily we are likely to perceive a new inspiring direction of social development. Yet understanding only the negative side of economic power will be of little value unless we can perceive the immense human and ecological potential of a different form of city.

Unlocking the creative spirit within cities requires no basic scientific breakthrough, mainly an imaginative vision upon ourselves as human beings. With vision of our specific inborn potential and the nurturing of social and environmental dynamics, we can develop a constitution for cities that will establish vast new dimensions for democracy, even as a dynamic democracy can itself promote a progressive urban structure. Cities and democracy are thus reciprocal.

Cities have failed in America because fundamentally we never cared about them. We gave them little insight or interest, continued to sentimentalize a vanishing rural life, asked only that they offer a good income, and never imagined that a grievious spectrum of crises would one day challenge their worth. Nor did cities have the pleasure of a theory by Darwin or Einstein, let alone one by a strident Marx or a deep-probing Freud.

Nevertheless, we did build cities with great energy and speed to unprecedented populations, and to what earlier observers would have considered unimaginable expanses of land. Never mind, the mood was confident of endless resources, endless growth. Spacious lands were cheap and ever speedy auto transport spanned the great distances. Challenges, we believed, would be met in the course of growth itself.

Cities, therefore, never found a clear public ideal, only a short-lived *City Beautiful* movement around 1900 and a few unnoticed new towns. But these efforts tell us as much about what was not idealized as about what was. Visions failed to appear for public environments, social development, cultural achievements, or even urban efficiency.

Without such visions, expediency prevailed, leaving the city a vast free playfield, just about the freest possible playfield for laissez faire competition. And what seemed most practical was auto-created suburban living. Sizes of houses and lots grew, as did the number of cars and distances traveled. The *Great American Commute* was the result, a daily ritual of morning implosions to downtowns and evening explosions to the suburbs. The commute was fully matched by The *Great American Strip* of boulevards with endless commerce, a hopeless irrationality of disconnected functions, and a visual cacophony. Shortly, commercial concentration began in single-developer shopping centers, and these centers completed domination of the city by the automobile, no less than the freeways racing to outer suburbia.

Early on, it mattered little that the act of a rural family moving to the city itself transformed life as much as did the products of the city. The new urban environments seemed to convert nearly all behavior into a system of exchange, with business hierarchies, paid entertainment, specialized repairmen, advertisements, even store bread. The cities to which people migrated were organized mostly for industry and product. The mass urban formlessness grew as a system of economics more than one of cities. But the unprecedented urban growth created a huge, complex, and interdependent mass, barely functional but lauded as the growing standard of living. That interdependence was a success of economics, not physical or social, so human isolation and alienation grew apace.

The wide band of unfocused human needs left unattended grew into deep problems spanning most of urban life: dire poverty, housing of utter inadequacy, gang culture, corrupt politics, fractured families, rampant environmentally based diseases, and personal alienation. Slums of grinding hopelessness were partly met in time, bit by bit, with programs of law and order, sanitation, welfare, unemployment benefits, housing, and urban renewal. Whether sweltering tenements harbored up-from-poverty success for the next generation or a festering stagnation into a culture of permanent defeat depended on the will and sacrifice of both individuals and families. Education, probably the best opportunity society might offer families to break away and avoid bitter endings, often became an exercise of desperation in which confrontations between students and teachers almost reduced success to maintaining classroom order.

Citymaking was a split affair, and no one, between the wide initiative of developers and the regulations of government, was clearly responsible. Many competing developers operated on the chance availability of lands and the fluctuations of housing markets. Municipal and county regulations had limited effect, yet public money had to follow the quixotic patterns of development with boulevards and freeways, water and sewers, schools, and parks at whatever the urgency and cost. Neither the pattern nor the process of development were controlled by anyone, even when the all-too-general plans were prepared and put into place. Not only did America build cities in naiveté and without either a base philosophy or a creative tradition of urban life, we built them with national prejudices going deep into our past and into our self-image as a people:

1. We *distrusted cities* even before the American Revolution. Cities were sources of evil in society, we believed, and no less than Thomas Jefferson warned of them.

2. *Pioneering and homesteading*, which challenged the vastness of great open spaces, still today remains solidly in our collective psyche, over a century after the frontier disappeared. So people continue to challenge all of the space they can afford and all of the distances they can drive—right in the city.

3. American *farmhouses* were built on each farm, in contrast to those in Europe, which are mostly concentrated in farm villages (that take on urban qualities even in the smallest clusters). However, Americans built their farmhouses in the city exactly as in the country, an ironic transfer of values symbolized a grave antipathy for the cities they built so vigorously.

4. The tradition of *rugged individualism* is easy for one to identify with while watching television from an armchair. But this myth of individualism runs counter to the realities of modern bureaucracy, technology, product culture, and especially media entertainment, and sportscasts, that is to say, to almost everything urban. That tradition seems to demand a detached, single family castle that one can command and effectively defend.

5. The companion tradition of *privacy* strengthens the role of the detached house and personal freewheeling auto, but too easily resolves itself as defensive isolation. The myth assures a distortion of both public and private values.

Taking these five observations together, one need not wonder why cities continue to fare poorly in the American practice. At best these deeply held notions implanted in the public mind a hazy and ambivalent view of what cities really are or might be. These pointers to our historic psyche explain why we cling to narrow material progress and fail to envision life in cities beyond personal wealth and power, or beyond the classic culture imported from Europe. The danger, I believe, is how virtuous traditions become misguiding myths. These myths have so very, very little to tell us about today's reality. They help us not at all in perceiving and interpreting history's most profound revolution: the industrialization and urbanization of human life.

Urban Revolution

> *When you operate in an overbuilt metropolis you*
> *have to hack your way through with a meat ax.*
>
> Robert Moses

> *More time and money is spent on automobiles*
> *than on children.*
>
> Richard Farson

From the beginning, America has consistently held back from including cities in the American dream. Even as millions of people migrated to them for opportunity after the 1850s, the varied anti-city myths flourished, often for self-fulfilling reasons.

The rough, rapacious manner in which logging, mining, and oil towns often grew explosively, declined, and died was often characterized by gambling, prostitution, and shooting as much as by what each produced. The filth and drudgery of river and seaport towns and sooty, gruesome steel and textile mills all added to the bleak image. Never mind that important inventions were being made or that a completely new foundation for life was being established. The work was dangerous and unhealthy. Child labor and sweatshops were used ruthlessly in pursuit of high profit. Housing was minimal and lacked fundamental sanitation in crowded conditions. Tenements for migrants without resource were built in endless, crowded rows with little light and air. If living conditions were as bad in rural areas, the vast teaming slums compounded the urban distress, as did crime.

In time, the expansion of cities out onto new lands with blocks and streets accommodated endless housing expanses and unknown future industrial and commercial needs for space. So the deathly stench of slaughterhouses, the ear-splitting noises of forges, the brawling streets, and sundry other insensibilities were randomly interwoven with houses until the expanding scale of industrial plants could no longer be accommodated on the few acres of a block, or until zoning in the twentieth century began segregating industry and housing—and promoting commuting.

Another strong reason for a weak urban tradition was simply the struggle to make do and to improve the worker's own lot. Slowly at first and by stages, families acquired indoor plumbing, hot water, central heating, gas or electric stoves, refrigerators, radios, TV sets, and automatic washers. Acquiring these appliances roughly paralleled improvements in family income. Here were improvements in the standard of living, and these were physical, close, and very real to the family.

Each appliance drew attention to the interior of the house, not to the urban condition outside. Most families had modest ambitions, and those were being fulfilled regularly inside. The outside urban environments mattered little, beyond a picnic in the park, a night on the town, and perhaps a double feature at the movies.

The American dream was thus working. It was an up-to-date myth with real substance and success. But its very success distracted attention from the issues that would in time grow into urban crises. Many grave problems had been resolved with new water and sewer systems, building codes, fire and police protection, and transit systems. Again, therefore, the general assumption was that problems could be resolved with increased wealth in the course of affairs. The attitude, consequently, was to confront one issue at a time, when each ballooned into a crisis that could no longer be ignored. However, the many special crises and the single-issue solutions designed to solve them reinforced the newly damaging myths of pioneering, farm homes, individualism, privacy—and the distrust of cities.

Slowly at first, starting after World War I, a new range of problems began to appear, and these were associated with wealth, not poverty. A growing numbers of automobiles appeared, and they began to alter the urban physiology, helping initially to reduce horse-drawn congestion, though soon creating their own. But through the 1920s their numbers increased dramatically, and a new kind of change appeared that eventually altered the very anatomy of cities, as well as their physiology.

It was in changing both the anatomy and physiology of cities that the automobile became revolutionary. Even the idea of the city changed. People individually and cities collectively came to understand the car as a fact of life, one that deserved its place in the city, and dominated it within decades. The city then had to be restructured, and the car was the tool and beneficiary that transformed all urban life. Never before had any product so altered cities like the automobile, or so abruptly in less than a generation. Just about everyone was captivated. Imagine: a Sunday drive in the country, a business trip to Chicago, a convenient way to get to work. Moreover, common wages were reaching a formerly uncommon capacity to buy a car. And industry was ready with its startlingly efficient moving assembly line. It had the impetus, ultimately far more than had the great railroad construction era from 1840 to 1900.

But the cities themselves were not ready. With weak traditions and a lack of visionary ideals to give them independent strength that could shape their own best form, they were helplessly open to the forces of automobility. Stoically, it seems, they went through one surgery after another to make nearly all bone, organ, and tissue of the city accommodate ever more automobiles.

The auto expansion started innocently enough by inducing modest improvements, such as road straightening and paving (getting out of the mud), then centerlines, stop signs, and signals. Soon boulevard widening began in earnest to reduce the traffic congestion (getting out of the muddle). Then the full power of automobiles began to be revealed. Their singular ability to alter the whole city was astonishing and radical when we examine its massive scale (compared to that of a person), its unprecedented speed, its parking requirements, and its numbers—all of which consumed enormous acreage.

When the automobile did appear en masse, it created immense pressures for space. But, far beyond its own demands for space, the automobile's role was unique—and revolutionary. *The automobile itself could make endless acreage easily available for urban development.* Any road into the country would do for a start.

However, the space it made available also was enough to endow another almost equally revolutionary force with renewed vigor, the rural-style frame house. Thus, with both house and car, new suburbia emerged. The new suburban lot accommodated a driveway (awaiting a second family car) and garage, along with front, side, and back yards. With its virtually unlimited ability to access the huge acreage, the car made such a large lot possible, then popular, and finally, almost necessary. Moving from its old central-city row house, the new middle-class family now purchased a house distinctly less urban. By comparison to what was left behind, that setting symbolized the tranquil and spacious countryside, even when it was a minimal tract house. This symbol buried itself deeply in the American mind, paradoxically reinforcing the fundamentally anti-city traditions, and it continues to expand in scope to this day, signaling the continuing rural identity of Americans.

Here we can see how strikingly the automobile, creator of great pressure for land, and taker of more land, completely dominated urban development since the 1920s. Along the way, cars prompted the innovations of supermarkets, shopping centers, motels, and varied drive-ins. Automobiles also developed auto sports, from four-wheeled adventures, dune buggies, stock car and drag racing, to the Indy 500. Today they carry skis, bicycles, surfboards, hang gliders or lug boats, camping trailers, sailplanes, mobile homes. Motor homes often pull a spare car.

Then, with greater distances to go, distances stretched transit apart, people increasingly commuted by car downtown to work and shop. This led to an entirely new conquest. The people came downtown in large numbers and demanded more than the pitifully few curb-parking spaces. And they got them, first through demolishing buildings for parking lots, then for parking garages. Here, again, the automobile demanded space and was the means to obtain that space, but this time right in the city's compact and formerly efficient downtown heart. The auto put a leash around the city, steadily tightened its grip, attacked the center and today commands the form, function, social relations, and the deeper psyche of the people.

The style of conquest of the city by the car proceeded like its reciprocating engine, into the country by the reaching and taking, then into the downtown by the same reaching and taking. Always its own needs for parking spaces, access streets, and boulevards were paramount for movement and parking at both ends of the trip. After first accessing the urban periphery and then infiltrating and dis-integrating the older metropolitan core over the decades, automobile commuting then became serious business. Most retail businesses of the old city had to find an entirely new validity by accommodating the car, or suffer a disastrous decline of driving customers. A special irony occurred when about 1913 Henry Ford first organized the production line and, as a consequence, *reduced* movement in his factories by a stunning ninety percent while his cars went into the city and helped *increase* movement by nearly as much.

While congestion was but momentarily eased with each new road or parking project, the congestion continued unabated and grew, which was lamented as the price for progress. Then the 1930s saw the first examples of a radical new scale of expenditures for automobility. The freeway idea was invented to be the ultimate roadway for motor vehicles. It encompassed four costly elements: multilane roads, median dividers, limited access, and grade-separated crossings and inter-changes. All were built precisely for the automobile's speed and performance. Prewar examples occurred on Long Island, in Los Angeles, and on the first stretch of the Pennsylvania Turnpike. After World War II, the demand for freeways exploded with new automobile production. Block wide swaths began to be bull-dozed through the living parts of hundreds of cities. Ultimate automobility finally arrived, and freeways themselves fully dominated citymaking from the 1950s. A complete pattern of urban freeways was substantially laid down in many cities by the 1980s. The new, auto-dominated urban anatomy was in place.

With improvements in trafficways, always one step at a time, every new step was never enough. The next steps to expand parking, surface-street capacity, and add freeway routes always turned out to be merely a phase to another new, more massive scale of expenditures on more fronts. But today the constant automobile penetration, constant congestion, and constant struggles to find the final relief have reached an impasse. But no relief is in sight. Only now is the social conster-nation and technical impasse of automotive transportation by one *improvement* at a time becoming apparent. The effect upon the city has become disastrous. Urban problems, we now know, were not and could not be resolved in the one-step-at-a-time pragmatic responses to growth.

The radical nature of what was simultaneously happening to the city through automobility was hardly recognized in 1950. Yet, ironically, in spite of the many urban agonies, there is hardly more awareness today. While disillusion prevails, popular comprehension is minimal. And so there is neither a Vietnam-style anti-war protest, despite far greater carnage and destruction, nor a conservation movement, despite the fact that cities are the heart of human ecology. The evident automotive necessity in cities, the driver's feeling of total freedom, the car's intimate luxury and owner's prestige, and the pragmatic responses to its sheer momentum have apparently blinded people to the disaster that automobiles have created for cities. It is, of course, a luxurious disaster for most people, but no less real for that fact.

Each time it seemed that proposed solutions would permanently relieve congestion. But a new scale of congestion was followed by new, seemingly final solutions, none more radical, costly—and illusory—than the urban freeway. Soon after they open, traffic usually overreaches planned capacity. And when congestion is encountered, that capacity is abruptly and drastically reduced. When in time the congestion spreads out beyond commuting hours, at no daytime hour can one be assured of avoiding freeway gridlock. Even eight-to-twelve-lane freeways end in frequent logjams. But as one might suppose, traffic volumes for the most part tend to flow just below the threshold of complete intolerability. Thus a total collapse never quite occurs. Nevertheless, in only a few decades, Los Angeles, Chicago, and even New York with its extensive subways, have so transformed themselves that they cannot function without freeways.

Now cities are finding a dead end for their freeways. They cannot overlay a new freeway pattern on top of their old one at escalating billions of dollars while chopping urban areas into ever smaller, ever less tolerable sections of inevitable neighborhood decline, while also dumping much heavier traffic burden on local streets and parking facilities. Here the blind application of technology drives government to do its dirty work, but operates without social vision or goals, but fulfills the growing escalation of requirements, accepting once again more automobiles.

In considering more urban freeways, society can no longer ignore their corollary impacts, the damaging *friendly fire* they have on every neighborhood that they cut through. Responsible concern for the city can no longer promote automotive transportation with clear conscience. The volatile conditions surrounding fossil-fuel depletion, fouled air, loss of urban livability, and seething social distress must now take center stage.

The automobile's revolutionary impact upon the city stemmed from four powerful combined forces: (a) First was the obvious demand for roadways, especially the boulevards and freeways, to assure *movement*. (b) Then it rearranged the city itself to give complete automotive *access* to every parcel and significant urban function. (c) Of course the automobile everywhere required *parking* in the most valued locations. Every car demands at least three parking spaces: one at home, one at work, and more than one collectively at shopping, church, hospital, school, stadium, and restaurant. And (d) the final revolutionary demand of the automobile was for the creation of its own commercial *services*, the necessary sales, repair, and fuel support. Today, even after a century of living with the car, it remains difficult to fathom how these radical forces could affect urban life so basically in so short a period of time, and so negatively for an otherwise rational society.

How could the country allow the automobile to become the absolute broker between the individual and all of the necessities and benefits of life? While the public believed that the automobile would improve access, as it did in the countryside, in reality the new auto-made urban anatomy has *isolated* the individual, including drivers but especially non-drivers.

When people become isolated in sprawled suburbia and forced to travel on congested boulevards and freeways, the diabolic impact of automobiles is tragically exposed. At horrendous costs to government and persons, and to the destruction of the traditionally coherent urban form, the gross effect of the automobile results in a *major loss of access* to the destination points of the city. That is, all access among destinations becomes increased burdens. The automobile has become the core force behind economic determination and the rapidly approaching economic monopoly over society. Economic logic thus becomes inverted to promote economic exchange. The losses to the automobile include money, environment, resources, usable land, human efficiency, social coherence, and cultural opportunity. Then traditional qualities inherent in historic urban form are utterly destroyed, including neighborhoods predating automobiles. Rational science, urban design, accounting and economics applied to cities have been suspended in the case of automobiles. Of course, each new parking garage or new freeway section opening seems to make good sense, like the relief resulting from removing a sliver from one's hand. However, important human values are lost to the internecine conflicts resulting from the overdeveloped, minimally effective, continuously congested, and wastefully consumptive trafficways.

Yet, having transformed the city for automobiles, society is locked into destructive automobile transportation. And because cars and roadways promote each other, people cannot conceive of other urban options, especially because the car's immediate availability and seeming conveniences give people today's greatest sense of personal empowerment. But the loss of access nevertheless signals the black hole of auto domination of cities.

And, while it was believed that the automobile would encourage and accommodate cities built to any size, as it did when the cars first brought urban development into the countryside, the public is finding that the limits to urban growth are imposed by the congested and now largely non-expandable freeway networks. So urban growth is beginning to focus on smaller, less impacted cities. While California has been worried for decades about water shortages placing limits to its growth, now suddenly it is facing a condition in which the limits to its growth are the mammoth demands for more automobility and the growing crises of just keeping *existing* volumes of traffic in motion. Therefore, the promise of the automobile has come full circle, and is reaching its antithesis, a defeat of automobility now set against the need for another expansion, now rapidly becoming impossible, which would make the city even more uninhabitable.

What is most astonishing in viewing the last century of urban development is the great shift in the nature of urban problems. Formerly, the acute urban issues arose mostly from general poverty of the people and involved basic questions like disease, water, and sanitation. But then, between 1920 and 1950, with some time out for depression and war, the problems took a radical new turn: they arose from wealth itself.

The focus of that wealth was, of course, upon the automobile and its suburban spread. Although expenditures never seemed enough at any one time, they did underwrite some momentary relief, but their greater effect was the voracious demands of automobility decade after decade to stretch the limits of both private and public purses. Only an unprecedented scale of capital could create the massive and dispersed suburbia, build costly boulevards, freeways, and parking structures, all while reducing the efficiency of transit, commerce, industry, utilities, government, and private citizens. Consequently, the city has become beholden to the automobile, where so much capital for urban formation has gone, and where so many modern crises have arisen. Wealth has become less the solution in the city and now is combining with automobility to become a basic source of its problems, all too tragically for the humanity of it all.

Against All Wealth

> *Scientific technology is presently taking modern civilization on a course that will be suicidal if it is not reversed in time. What, for example, will be the ultimate consequence...[of] the overwhelming majority of urban dwellers [desiring] to occupy a one-family house, to drive to work in a*

> *private automobile, and to identify leisure time
> with essentially aimless movement?*
>
> Rene Dubos

The model for suburban America resolved itself into a single-family house and yard served by two or three mandatory single-family cars to maintain vital connections to three, four, or five urban districts to which the family must be engaged throughout large metropolitan areas. The family may live in one jurisdiction, shop in one or two others, work in two more, and attend school and church in others. Little activity is close to home, virtually none by walking. Civic virtue, and responsible citizenship as a consequence, have become casualties of multi-jurisdictional involvements. Home becomes but a command post with daily R&R of bed and breakfast. Even vehicle repair poses a logistical and financial crisis.

But then we discover how suburbia fundamentally isolates the family. This is acute for children needing playmates. Playing Little League is usually possible only with full engagement of each family's transit system. The tranquilized isolation would do justice to the Swiss Family Robinson were it not for the constant, frantic motorized commuting.

Thus we see how the two dominant urban forces of single-family house and single-family car, basic facts of the American dream, pose the ultimate conundrum of modern American life: the city that is anti-city; a mobility that threatens to become immobility; a house that relates to no urban or rural activity; miserly, little-used private minispaces that destroy the possibility of larger, more varied natural areas; an environmental consumption threatening ecological disaster; a standard of living that denies a quality of life, a luxury to sit in dead-end congestion; a material wealth that must continually increase to avoid social poverty.

Still the single-family house and car together have formed a withering vortex accelerating the great wheel of production and consumption that propels unending economic growth. Its dangers, wastes, tensions, and lost time are created and organized to promote that growth. The wasted resources would make many poor nations wealthy.

Meanwhile, the auto's explosive impact on urban space—and its infamous manacle on public money—has occurred with a traffic system that works well only at low traffic volumes and can come to an absolute halt when congested. Ironically, while great car-operating costs and exorbitant freeway costs create a system of shockingly low capacity, the powerful freeway lobbies (auto, gas, construction, trucking, repair, motels, and restaurants) promote more freeway construction, which destroys public transportation policy as much as it's continuing urban fragmentation. But today, under the auto necessity that dominates urban form, everyone, it seems, supports more auto transportation: families, corporations, municipalities, states, and federal government. And they must, for nearly all urban environments are now absolutely in the dependency of the automobile, a solidly drug-like dependency. Sadly, one-step, specialized pragmatism in fighting congestion remains paramount in public thought. The car in America today holds a fatal combination: It is lifeblood more essential to economic stability than defense industries. It has become a social necessity, even required as much as a place to live for those below the poverty line. It requires the attention of a disease, being like crippling polio to owners, a dangerous HIV epidemic to the public, and a very profitable care practice to industry. Non-drivers lose jobs, memberships, and their effective citizenship.

While the automobile dominates a city by radical and random dispersion, which makes it the necessity it is, public transit must operate on the opposite principle of serving compact urban development. Large volume usage is necessary to make a go of it. When the city is dispersed over vast expanses and urban centrality is lost, an impossibly huge transit network would be required for an impossibly small volume of passengers. Public decisions were therefore insidiously preempted decades before the public was forced to think seriously about transit development.

The auto-transit dilemma is best illustrated, of course, by Los Angeles. Between 1945 and 1965 that city killed its excellent regional Pacific Electric network, then repeatedly voted against a new subway system until the late 1980s when it was forced by freeway gridlock to start construction on a system that lost three or four of the most critical decades of influencing the pattern of new development in a positive way. Thus, finally, the city built by freeways is forced to recognize the deeper imperative of urban form, even as the first transit lines will serve mainly to protect some freeways by relieving them of enough congestion to maintain minimum movement.

Yet even this fundamental lesson has not been learned. When the automobile dominates the city and the political lobbies are able to effectively deprive transit funds in favor of *critically needed* freeways, an irrationality of urban development prevails and more automobility is assured. Then freeways cut through urban flesh to give brief congestion relief at an entirely new order of cost. Should some freeways not be immediately congested, they encourage more people to drive longer distances in more directions to more rural-style houses up to each family's financial limits. The city has become a never-ending enterprise to maintain motion. But now, as freeway networks are mostly completed and the insatiable auto continues to create congestion, the freeway gridlock inevitably demonstrates its ultimate impasse and certain disaster if present support for automobiles is further pursued.

Yet, implausibly, such endlessly expanded mobility remains an ideal without any serious question of its ultimate effects: on resources and environments, on simply urban livability and function, on both public and private finance, and upon social insanity—that very silent but frantic disease robbing the public of soul.

Then the irrationality becomes a paradox. The modern metropolis, with lower and lower population densities of development, exhibits greater and greater congestion. Low density of population, not high density, we are now sadly learning from the effects of the automobile, creates modern congestion. Inevitably, then, the mammoth expenditures to serve the scattered urbanism have a decreasing ability to overcome the aggravated congestion. That is not just the lesson of Los Angeles, but all cities over 50,000.

America has exhibited two major forms of the city. The first bustling city of the late 1800s was underdeveloped in poverty and minimally functional. The second city of today is overdeveloped in wealth and also minimally functional, overrun by the tragic technical trivia dominating our age.

The importance of the two cities today is that both represent the long struggle for a better life through product development, notably again, the house and car. That was the American Dream. People in the earlier city, with little material comfort, had a broad hope. People in today's city, with material comfort, have splintered hopes and grinding material contradictions. Where is the new vision and new hope? Where do we go from here? Should we, can we, Los-Angelize the continent?

And are not huge, older areas of today's city—despite all modern wealth—falling into collapse? Even if we don't comprehend it, we see it: ghostly buildings, drug alleys, speechless streets, terrified businesses, dead cars, sidewalk trash, sleep-trap motels, and schools displaying their learning in graffiti. Or is this but an apparition from 1890? No, the older deprived city continues in the dilapidated areas around the downtown left behind by those who can afford to escape into the urban sprawl.

Here are the two sides of the modern spirit, one finding defeat in the suburban distractions of wealth, the other facing defeat in the decayed environment that was left behind for defeat. Here are the two sides of the modern spirit, both poisoned by a noxious inferno burning on urban materialism.

City as Product

> *Most of man's problems [arise from] the stimuli of*
> *urban and industrial civilization...They are not*
> *inherent in man's nature but are the products of*

his responses to social and technological innova-tions.

Rene Dubos

I've heard it said that the American city lacks everything from which no profit is possible.

Wilfred Owen

Cities not only produce. They are themselves a product—a very complex product that deeply affects what we are and what we do every hour of every day. The effects burrow into our personal behavior, quality of life, work and commuting, finances and expendable income, government budgets and taxation, enterprise efficiency and competitiveness, health and safety, crime and punishment. All these gigantic results stem from the manner in which the city is created and how this master product of society will affect people's lives for many generations. Here we can describe the processes only in very broad strokes.

Municipalities have struggled to improve the processes of urban develop-ment—except roadways—largely through zoning and subdivision controls. These controls segregate basic activities and set standards of development for any parcel of land close to the city—and not so near the city after World War II. The land parcel's size and shape, established by the random forces of real estate, con-found prospective urban layout.

Municipalities also create plans to align zoning and subdivision regulations with transport, utilities, schools, and parks. These plans are hardly approved when state highways adopt a conflicting freeway alignment or when school sites are not purchased in time to save a planned location. Thus traditional land tenure and conflicting governmental practices both are superior to and undermine the plans, almost before they are initiated. Planned unit development, councils of governments, and local agency formation commissions have helped strengthen the planning process. But the fundamental difficulties remain.

Plans that attempt to limit distant urban growth are rarely effective. Dispersed housing tracts often appear in county areas beyond municipal con-trol, undermining and breaking up the official plans for internal urban devel-opment. With highway development underway, large shopping-center developers then vie to position huge shopping malls at points of greatest traffic (not greatest population, one must add). Each new mall forces a sales decline downtown, making it less viable as a multipurpose heart of the metropolis. Further escape from older, close-in housing is accelerated, leading in time to slums or abandonment with predictable consequences. Highly capitalized corporations regularly obtain major zoning variances and special-use permits to build their shopping malls on sites they choose, the urban plans be damned. The stakes are high. To develop as they wish, they mount lobbying and publicity campaigns, threaten local politicians with election opposition and cities with lawsuits.

The costly maneuvers further destroy major parts of a city's plan. One of many dilemmas for planners is that wherever they might propose a large and very crucial shopping center, they inevitably offer a huge windfall to those property owners and a windfall denied to others. So inevitably, urban development remains weak because local government does not have the *organizational initiative* to create more unified and imaginative plans and make them minimally workable. *Short-term private profits win over long-term public purposes.* The public interest is distorted. The urban potential is lost when the city is made the playing field for organizations with only their special interests being calculated.

Housing itself is but a class determined commodity bereft of social and cultural possibility. By design, the single-family house is isolated from commercial services, schools, cultural facilities, parks and recreation, and even sociable neighbors—enforced by fences, street space, and time-demanding distances. Then, paradoxically, the automobile further reduces overall accessibility in another classic case where *tactical success*—and necessity—favors the car while there is a *strategic* failure to establish the natural and historic efficiencies and amenities of cities. It is time to put to rest the barren and self-defeating concept of housing, along with numerous other destructive urban development practices. We have yet to conceive the family residence as a vital cell of a complete urban organ like the body's heart and lungs.

While the corporate interests have the initiative to plan, develop, and take profits, public authorities are then burdened to follow up with highways, schools, parks, libraries. By this process, the form of the city becomes more chaotic, degenerative, ecologically destructive, socially demoralizing, and especially costly. Public resources must be stretched continually in the struggle to serve the unpredictable and chaotic growth. The land developers, having the critical urban-development initiative, leave behind an urban formless subdivisions no one alone would devise. Developers can then depart from their subdivisions while the public authorities must tax and bond heavily to follow through with urban infrastructure, rarely to common public satisfaction, much less inspired achievements.

Schools are hit hard. Administrators cannot judge whether to build schools to the west or the east, and whichever way they go there is at least some error, sometimes a major waste of public dollars in underused or undersized schools. They find it difficult to plan or purchase sites because they cannot predict lot sizes, population density, and therefore, reasonable school-service areas. Consequently, very costly urban school bussing must be extended to more residential areas.

The wild second guessing the suburban development locations, directions, and rates of growth also applies to water, sewer, gas, electricity, and telephone utilities. Generally, the utilities must be planned for maximum possible, or excess, capacities, which must then be charged to the public in higher utility rates. Other public costs increase in the dispersed, low-density suburbs. Mail delivery must be motorized with greater costs and reduced efficiencies. Police and fire protection are similarly extended and less effective.

Most urban plans thus remain largely confused or sterile, having little impact on key issues of either declining central area or peripheral scattering, let alone any possibility of imaginative development equal to our modern potential. Principal power resides with capital, not regulation, so the open urban playing field of development is dominated by business developers followed by public road construction. It is doubtful whether regulation can ever result in sound area wide design. Land ownership and competitive real-estate maneuvers clearly carry greater public power. Yet the results will remain many decades, even centuries, that the occupants must live with the resulting ill-formed urban areas.

We have, then, a *chase* system of urban development in which profit-making subdivisions based on the vagaries and tactics of the real-estate market force large-scale public finances to follow suit, to build the urban infrastructure after the fact of private development, assuring a lower efficiency of municipal service.

Then there is the old city left behind. Both private and public moneys are effectively denied to improve older city areas—which continue to fester and generate more costly public problems. These accumulating problems grow into crises. Without extraordinary public incentives, profit-making organizations will not touch areas of rundown housing, dispersed ownership of small lots, existing streets, and buried utility lines. Public authorities consequently have little power in the older areas to stop their decline, let alone upgrade, renew, or reform.

We see, then, how land speculators and private developers force the public purse and decision making. Rationality of urban form steadily breaks down, which is overcome by more costly solutions of transportation, utilities, and schools. More traffic must move greater distances in more directions. The anti-city therefore becomes central to the citymaking process.

The chase form of urban development creates a pattern of injustice. While great profit occurs in converting rural land to urban uses, the new residents who create the demand for residences underlying great increases in land values do not share in the windfall profits of the land. Nor do cities, counties, or school districts benefit, although they are forced into inherently chaotic, costly, and mandatory investments and services. Then, simultaneously, they inherit the anguishing problems of decaying older areas left behind by private capital fleeing to profitable suburban development.

With the explosion of urban space to serve automobiles and large-lot houses, the automobile has become an absolute but deceptively destructive party to urban life. This makes a farce of the point of law that driving a car is a privilege (and, by implication, only for pleasure). Plainly, then, the young, the elderly, the poor, the handicapped, and all who cannot drive are deprived of citizenship. Since the car has sent public transit into deathly decline in most cities, all people denied the new auto necessity are deprived, or live at the sufferance of a driver. Agencies who help the non-driver also must submit to expensive automobility: school bussing, mobile libraries, meals on wheels, and special transit for the elderly and encumbered.

The irony of urban automobility is that it remains virtually sacrosanct in public discourse. *Why* is a question rarely asked. Only the question *how* to accommodate more is debated. Even those people who fume about stop-go gridlock on gold-plated freeways rarely broach the deeper questions of the auto's debilitating effects on the city and on every person.

Americans have sought mobility, and they have achieved mobility most poignantly in the cities where it is—both theoretically and practically—least needed. They have made massive mobility a damning necessity when they might have made moderate mobility pleasurable, especially when access could be improved solely through urban design.

At a time when the world strains to face massive social dislocations and environmental problems, from global warming to the decline of species, the automobile escapes serious censure. Many people blame population growth, not consumptive auto-ridden cities—a bad misjudgment by those working on a valid but virtually separate issue.

Long ago we should have recognized that nothing—repeat, nothing—consumes, pollutes, wastes, and destroys more than automobility and urbanism cast over the countryside. Tobacco is blamed for more deaths. But cities constitute the whole living environment for most people and nearly the whole content of their entire lives. Their future is at stake. There is no exaggeration in saying that our humanism, as well as our human ecology, rests in the form of cities. Human alienation, environmental degradation, and the world's resources are too demanding for us to miss the common cause.

Considered as a whole, citymaking today resembles gambling at great stakes. This urban gambling table plays upon land, location, price, and building styles. But the gambling also includes maneuvers and deception, as does war. Whether the image is gambling or war, the results are made substantially by chance, suffered for generations. The simple tragedy of this open playfield of citymaking is

that any ideal of what a city might be is impossible to conceive, aside from advertisements of luxurious living. Subdivisions at best mean antiseptic bedroom
developments almost completely barren of social and cultural amenities (excluding country clubs at the highest level of suburban development).

The American public has yet to perceive that the rational, economic, ecological, and cultural validity of cities is precluded by the exchange system of creating
urban environments. The illogical patterns created are radically inefficient for
individuals, families, production and service organizations, and all levels of government. They are set indelibly in asphalt, concrete, pipe, and lumber, usually for
centuries.

Characteristically, cities are dominated by economics to the point of exchange being
virtually their end purpose, thus massively usurping human sovereignty. We need
only observe again that the more wasteful and internally destructive the city
becomes, the more it *contributes* to the Gross Domestic Product and the standard
of living. Consider the economic *growth* of an accident that destroys a car and
puts the driver in the hospital for a month. Or consider the growth involved in
building a utility system for one hundred million dollars, which with other urban
design might have cost less than half as much. There is hardly a subject of urban
life in which such ironies do not abound.

Obviously, the automobile is here to stay and is basic to a kind of personal
empowerment in this still-adolescent period of the technological society. The
automobile is indeed fine when used in moderation. Indeed, its only pleasure is
in moderation. Congestion displeases everyone. So it is the overdriving *necessity*
of automobility that must be eliminated. Can we ever stop the momentum in
cities to forcibly travel ever greater distances in more directions more frequently?

Suburbs today absolutely *demand* two or three family cars for family viability.
By distorted reasoning, prestige is attached to this fact, just as it is added to the
GDP. A car purchased and used like a yacht for pleasure certainly adds to a quality of life. But second and third cars purchased by necessity of commuting, shopping, and family chauffeuring are a critical loss to the family and a sorrowful
waste to society, which must build destructive roadways to accommodate the
family's arbitrarily created travel necessities.

A special irony confronts us. While our economic system diligently struggles
to reduce costs within manufacturing plants, the products created in those plants
are put to use in cities in ways that promote ever more wasteful growth. Socially
speaking, efficient production is defeated by inefficient use. Fleets of little-used
yard equipment, for example, are made mandatory by individual ownership of
large, but also little used, single-family yards.

Yet suburbs are peculiarly idealized as demonstrations of prestige as having fulfilled the American Dream. That dream, however, is accompanied by gigantic,

hardly manageable garbage dumps, huge automobile wrecking yards, and expanding prisons, all of which are terrible costs and expendables of industrial society.

Destroying Land and Urbanity

> *Our society offers no economic incentive to conserve anything.*

Kenneth E. F. Watt

> *Real estate operations require the subdivision of the unbroken surface of the land...into a collection of alienable commodities....This commodity consciousness affects our entire psychology about our landscape, atomizes it, puts merchandising spectacles over our eyes, makes it difficult or impossible for us to see the landscape as one unified and controllable experience.*

Garrett Eckbo

When the city itself is built as a product, the necessary foundation for that product is land, no less than concrete or lumber. In making cities as we do, based upon micro-parceling of land, with land-hungry, single-family houses and single-family cars, we maximize the consumption of land—but, paradoxically, not the spaciousness of the city. The consequences of this method of urban development have not been assessed or made part of the public debate, except possibly for the incidental loss of wildlife habitat. Here it is only possible to review a few salient points.

In building cities for supposed spaciousness, that is, by developing ever larger lots with a tight cobweb of roadways, we have not asked what this means for spaciousness, or for the many sectors of urban efficiency, let alone the qualities of urbanity and good living. Could it be, in this age of science, management, accounting, and unlimited analysis, that we do not know the varied and very real consequences of our aims and our methods in building cities? Could we, as I argue, be mandating a gluttony of land in the form of mutually exclusive mini-spaces, with ever-more oppressive social isolation? Are we not thus also requiring a gluttony of machines to manage and span those spaces? And finally, aren't the compounding gluttonies threatening the functioning and livability, even the survivability, of our cities?

Then, with the buildup of urban-land consumptiveness, substituting for natural space, we unnecessarily withdraw and consume millions of acres of farms, the countryside of nature, the watersheds and forests, and the habitats for wildlife. In so doing, we make scarce the nature in nearby rural areas, reduce the accessibility of what remains, and create neither spaciousness nor efficiency in the city. Every part of the process works against us and compounds itself to narrow the vitality of persons.

The great continental spaciousness of the American heritage is critically diminished. Are we not making a spacious land small by the incessant parceling and mutual exclusiveness with all our power? Are we yet to reduce our usable continental land area effectively down to the size of Great Britain while expanding the urban roadway pattern to the continent?

Historically, land was the major foundation of wealth. But industrialization changed that, making it into merely one more resource for development—especially to build the city as product. Land is one of the most finite resources on earth. Unfortunately, the current method of developing cities and using land are both destructive, both formed by the current models of economics as the materials for production and consumption, no less than the woods, fossil fuels, and crops. While cities today are built to be enormous consumers of land, the excessive infusions simply confine or destroy urban behavior, efficiency, amenity, and sociability.

Unlike most other resources, urban land is valuable only through the benefits of location or access, those reflecting the benefits or burdens of distance. The effect of the automobile has multiplied the scale in which access occurs, creating the illusion of efficiency and freedom. But in reality, access and location have become beholden to the automobile while escalating land consumption and destroying all semblance of proximity and human scale, and therefore greatly expanding distance determined inefficiencies.

More importantly, the central fact is that in becoming urban, land has been chopped into many thousands of minuscule properties. These parcels are exclusive to each owner. Therefore, urban public lands—the commons— and the spaces into which people can commonly and freely use account for only a negligible portion of the city. And most public land is committed to roadways and parking. The preponderance of urban land, being private and exclusive, is structurally anti-social, open only by invitation by owners for special purposes. Land possession thus rules behavior. However, cities are best when they consist of hundreds of common accessible social spaces, freely available for many purposes at anyone's behest. But today most of what we think of as public space is actually public only for the interest of the owners, that is, the spaces are the stores and shopping malls managed strictly for trade. They close at the end of the shopping hours, unlike true public spaces.

Thus, when most land becomes urban, it is exclusive to owners and organized to determine behavior, namely to purchase. The endless commercial or private parcels, like all environments, do powerfully shape human behavior: the patterns of work and shopping, making friends and pursuing many kinds of recreation, attending school and church. So today urban behavior responds entirely to owner's or renter's possessory interest. But possession runs counter to the best purposes of the city, that is, to freely afford diverse activities and associations. Thus private possession is fundamentally anti-social by both structure and the owner's limited interest.

The critical fact is that the pattern of urban land tenure and its reflection in transportation is imprinted upon the mind, especially since the pattern of land parcels governs the pattern of behavior. So today behavior and thought is narrowed by property ownership and heavily oriented toward commercial transactions. Unlike the Spanish plaza tradition, we don't usually have an urban center, a space belonging to everyone, one widely used throughout the day where many people congregate especially for evenings, Sundays, festivals, and celebrations. So we must ask, in what public spaces do Americans celebrate? Where are the places to go spontaneously, or on holidays, or regularly to be social?

Consequently, the splintered and possessive nature of urban parcels not only imprints itself on the mind, it restricts the person's sense of appropriate behavior and assumes that seeking the better life is an isolated and individual affair, sought mainly in commercial entertainment. Hence the mind views the fractured urban environment as if it is an unchangeable fact of commerce, however confining that image may affect our outlook and behavior. In fracturing thought, the fractional urban parceling limits one's vision to what can happen on a particular piece of land, and fails to perceive the unified social and cultural potential of a whole district or city.

As a result, residents individually hoard their endless citywide monotonies of square feet in the family's private open spaces, and thereby deny themselves the far greater urban potential, were there created diversities and far more public open spaces. But in today's vision and behavior, people strive to build private pleasures of the good life, sadly and overwhelmingly through television entertainment in a deadening passive mode. They thus deny themselves the far greater potential diversities of behaviors that arise out of a person's own social initiative, which are best or perhaps only expressed in common public spaces, and covering the range of indoor and outdoor interests, whether social, cultural, recreational, or simply public relaxation in lively natural or social settings.

The thousands of nearly identical and repetitive personal mini-spaces throughout the city demonstrate how individual ownership imposes an overwhelming uniformity upon social life and denies the individualism that is at the heart of the American Dream. Uniform and fragmentary existence is but another term for the urban masses, regardless of what class to which families may belong. Existing essentially alone as individuals, people are open to the persuasions and exploitations that all mass markets strive to establish. The consumptive monotony established by the house and car thus precludes development of broad human choices and opportunities that all cities should offer their citizens. Isolation individuates persons, denies true individuality, leaving them at the mercy of processing and promotions by the market forces.

While society has built a massive cooperative system of enterprises that produce a monumental range of products, it has not yet learned to build a comparable system of interpersonal cooperation and association, and with that diversity an equal variety of urban environments and facilities—conditions that help us to deeply experience and enjoy the wider possibilities of our historic wealth. The great range of potential urban choices is all but excluded by the monopoly of land consumed by, and mostly wasted in, the dead spaces around every house and by the huge spaces dedicated to mandatory but meaningless movement on roadways. These socially forbidding spaces constitute the largest spaces we have.

The potential of cities, and the foundation for the expansion of human experience, is shaped through the organization of land. Unfortunately—one should say, tragically—most of the dynamics of urban land is lost by the fractured monopoly of mostly useless or unused private parcels. These dynamics are also lost by the blocks and streets that systematically create barrier and throughout the city and segregate activities into sterile monotonous zones, which then create the great distances between the things we want in the city.

The benefits of public lands, however, can be derived from designed diversities, created for the comprehensive range of public interests. They emphasize facilities and programs, involvements and opportunities, rather than products, and consequently, doing rather than consuming. They offer many forms of association and cooperation, encouraging people to come together in many different ways for many kinds of activities.

But today, without a physical basis to found and expand the urban imagination, visions of public life have little or no basis on which to germinate and grow. So the mind is confined to the small niches of private lots, blocks, and their restrictive power over the mind. Broader visions are difficult to conceive and virtually impossible to develop. Socially, we are intellectually stigmatized and psychologically imprinted by the absolute fragmentation. Considering our vast individual and collective wealth, the cities we build critically stunt our human development. We are therefore losing the larger possibilities of our age.

Today's system of urban land is founded upon a chess game of controlled chaos in the real estate and urban-development market. So our urban vision is one of personal exclusivity. Beyond such facilities as a public arena and convention center, possibly an art gallery and museum, offer little more. So the bought-and-sold city establishes isolation compelled by the monopolized urban mini-spaces.

The controlled human chaos contrasts sharply with the unity, integrity, and efficiency evident in industrial plants. Inside, one finds a long tradition of perfecting the processes leading to an efficiency derived through the complete spatial organization and integrity of the production processes. Industrial efficiency is possible because the conditions of production are precisely and completely controlled. But outside, one sees how products and workers enter into environments of disorganized and deprived behaviors. There is little tradition of local or neighborhood social cooperation, and therefore little social integrity, and little thought of achieving greater access and public and social efficiency. The chaos of movement, the *high standards* of private parcel size, and the general promotion of wasteful consumption remain the order of thought.

The initiatives of industry are corporate, taken for corporate profit. The initiatives producing new urbanism also are corporate and they, too, are designed for profit. Those processes create a city and leave it for the public sector to build and maintain a public infrastructure. But they are founded on a primitive rationality, the rationality limited to real estate exchange. While a very high level of purpose, integrity, and efficiency are evident in industry, a barely maintainable order prevails in the city, a root contradiction we take for granted and do not explore.

Thus, while our present diversity of product and service stems from a magnificent tradition of varied and vigorous enterprise, the potential diversity of urban environments necessarily stems from a single and finite resource, the land, but is now organized solely for real estate profits. Fracturing the land does not lead to a meaningful and beneficial social diversity, but to its opposite, a banal social monotony. Hence, the route to a truly meaningful diversity of urban land—and its corollary, social diversity—requires a unified source of vision, design, and management. Arising first through design, diversity thus reveals a paradox, one that must be resolved for a greater range of urban freedom to occur and perhaps for a grand renaissance of cities to arise. But one can hardly imagine a human renaissance appearing in the sterile monotony of suburbia.

Where there is chaos, especially a *system* of chaos as the histories of fundamental human disruptions found in real estate development demonstrate, there is open opportunity for broad social exploitation. Not illegal, not named, not debated, and not evident in particular transactions, this exploitation nevertheless severely oppresses the urban potential. Everyone is deprived because everyone's life today is severely confined to favor economically sacred profits and our strong tradition of seeking material possessions, especially as against a philosophy of living fully and vigorously with our fellows in all departments of life.

Sorrowfully, the long and continuing history of chaos in citymaking blurs its consequences. Possibly the most serious is the urban deterioration that now grows like a scourge in large areas of metropolitan regions, sometimes in a period as short as twenty years after development. The layout, impact of traffic, suburban escape, and other factors propel the decline. Then the local economy of a district deteriorates, and social instability and crime accelerate the decline.

At this point, the greater consequences of the real estate market become clear: The small lots and their costly streets and utilities establish a condition that is virtually incapable of stimulating a valid second or third generation of improvement. Only complete redevelopment—wiping the land clean—can substantially remedy the deep structural shortcomings of the original development. But redevelopment is outrageously costly merely to reclaim land and becomes completely impractical when the automobile can reach so much more land distantly at a fraction of the cost. But that escape is precisely the force that causes financial abandonment of old areas and condemns them—and their residential areas—to decay. The painful process of redevelopment sometimes must deal with small lots, for with only 5,000 square feet per lot, eight lots render but one acre. Even minor redevelopment is of necessity public and depends upon tax money. But since very little redevelopment can occur, the results are both costly and feeble. Yet urban development at the fringes continues the destructive fracturing of land and results in essentially the same parcels that are at the root of the central urban decline and defeat. And so there can be little faith in the ability of private or public renewal processes to brighten the future of most old urban areas, even when considering some exceptions, such as Georgetown in Washington, D.C.

Land is a non-expandable, rigidly local resource. Its crucial role in cities is formative, shaping our behavior and thought for good or ill—far into the future. We cannot abandon it like an old hat or car, however antiquated, for it will continue to shape urban behavior whatever we do. We need, therefore, to better understand its dynamics and to plan it to reach the outer limits of the urban possibility.

Counterfeit Progress

> *It is rarely recognized that the slowest method of travel is often to use the fastest type of vehicle.*

Kenneth E. F. Watt

> *We must bring our power systems and our value systems into balance again.*

John S. Lambert

Every car itself makes vast demands for land. First, each car requires between three and four parking spaces, at home, work, and numerous other destinations. Second, automobiles demand land for the vast network of local streets, wide boulevards, and freeways, the first to maneuver in and out of driveways at each property, the second to collect and distribute local and freeway traffic and to access shopping centers, and the third to span the great urban distances generated by the automobile itself.

Over half of the land in many high-value but now-depleted downtown areas is devoted to streets, alleys, freeways, and parking lots or garages. Shopping centers devote up to eighty percent of their space to automobiles. Parks, zoos, and museums also must allocate increasing amounts of their money and land to parking.

These reciprocating actions that escalate mobility have been self-defeating from the earliest days of the automobile. But the increased mobility that was thought to be an improvement in access in the early part of this century has, at a fantastic cost and tragic reversal, ended with seriously reduced accessibility. Today the dispersed city is muscle-bound in a huge road structure of concrete and asphalt that allows no realistic alternative to the automobile. Yet in consternation, while there is no practical meaning, sense, or value, expanding the urban motorways and motor spaces continue unabated. This can happen only because the processes underlying automobility are buried drug-like in all thought and public policy.

Even a tally of direct costs of automobility are shocking, but they are accepted because the auto necessity is seemingly as entrenched as the law of gravity, and because the costs are dispersed and often hidden in many private, corporate, and public accounts. The public pays for freeway, highway, and boulevard construction, repair, and rebuilding. Developers pay for local streets and some boulevard construction, and these costs are included in lot and house prices. The individual, of course, must make car payments, while also paying for insurance, gas, and oil, small and large repairs, tires and batteries, parking and tolls. Each item is paid at different times and in different ways, which obscures the cost burdens.

For all of its pervasive power in being the ligament between citizens and all of their social privileges, the car's role has carried with it a very peculiar twist. When fast motorcars were introduced, one might have assumed that farmers would move to towns and cities for the many advantages and amenities offered there. Instead, in an ironic turn of the rural idyllic, they remained rural while urban people used their cars to move not only into the expanding suburbia but beyond into rural acreages of anti-urban isolation. The psychosocial dynamics of such an apparent social denial in such logistically difficult circumstances is indeed difficult to fathom. And so the term *sub-urb* is well placed. The average citizen's only conflict with the law is via the automobile, whether moving or parking. And it is the citizen's most lethal instrument. The automobile's record of fatalities far, far exceeds battle deaths in all of America's wars, seriously injures millions more, and continues year after year, but without a declaration of war.

What are some of the social lessons of automotive dominance in cities? Imagine each of the rooms of a house created as separated buildings on a lot. Consider the costs of construction, including structures, plumbing and wiring, heating, and air conditioning. Consider the yard, its functions and landscaping. Then imagine the effects on the movements and efficiencies of each person of the family in their daily routines among the dispersed rooms. Finally, think of the effects of this dispersion on the interactions and unity of the family, including, first, the simple conversations, then the effects upon spontaneous comments, jokes, retorts, bantering, laughter. What about being in contact during meal preparation, helping with homework, planning family events, and giving advice to children? While nothing is completely prevented from happening, family activities and interactions deny so much that is so spontaneously human from happening. Here we see a spatial pattern of alienation in which each member is isolated to its own functions, except when a specific family event is organized and convened. Fortunately, most dwellings have a unified room arrangement in which individuals may be a part of the whole family by sight or sound while they pursue their own interests, or close a door for privacy.

However, the city is not conceived as anything like the unity of a house. Functions are distantly and randomly separated and require conscious intent, purposeful planning and an organized trip. Not only are serendipity and spontaneity lost, freedom of action is lost to the trip and the work surrounding it. Consequently, associations are often reduced to contacts, and contacts are minimal. Plainly alienation exists, physically if not socially and psychologically. Still, in the modern ethos, such an existential alienation is counted as a positive expression of urban life.

In the many interrelations that constitute the American city, transportation has clearly become the overwhelming feature, defeating the concept of proximity by which all cities have demonstrated throughout history. While we imagine transportation today as bringing us together, in the reality of cities the major expenditures, hassles, and time required to get together are solid measures of personal isolation, social fracturing, and human alienation. We must always wonder if a trip is worth it—and then decline to take burdensome trips unless they stand high in importance. How many trips we pass up without a second thought may be a more accurate measure of the alienation. The less burdensome the trip, the more likely we are to take it, which is the lesson we should learn in planning cities, that is, to promote urban *access* rather than transportation. Here is the lesson of the psychologically debilitating human estrangements that are so pervasive in cities today.

What a good city needs is immediate or very close access between many kinds of things, like a unified house. Access in a city, when planned in the manner of a house, will not only reduce movement and its many severe costs, it can simultaneously increase the free, varied, and meaningful associations among people. Close access can be achieved only by proximity and the integrative design of urban spaces and functions.

We must ask again whether a city built for escape can avoid disaster—functionally, ecologically, and socially. Can the spatial pattern of a city defy proximity and access, that is, can the pattern of the city defy laws of space and distance? Can it do so without exhausting itself to overcome the anti-city expanses?

We know too well how single-purpose decisions works against human progress. Yet the automobile has also created a single dimension—increased mobility—in which decisions affecting the city are made, as when gas taxes serve only automobility. We call it monopoly and forbid it in the conduct of business. Imperial communism was perhaps the world's greatest monopoly, resulting in one of the world's greater tragedies. Now we see how an industry, that of the automobile, has become a new monopoly dominating the whole of citymaking. It has become the monopoly of urgent concern, and its effects are killing, enervating, and depriving while being disastrously costly and wasteful.

The automobile, being the greatest single result of industrialization, has become the greatest mistake of industrial society. It has become the dystopia of cities and the principal reason why the cities are ecologically unsustainable. More than any other cause, the car has so deformed cities that the conditions of a social meltdown are appearing. They consist of upward-mobile persons facing stunted affluence, alienated personalities dominated by mass media, and all who find intense frustrations in their decaying urban environments.

The lessons of the automobile dominating cities are buried deeply within us. Today this irony in American industry and affluence contradicts basic lessons of urban development. In the distant past, cities were founded as temporary encampments, gradually attained permanence, then sometimes achieved a degree of amenity. In some cases, greatness was achieved, as in many Roman cities. American cities, however, confound the process through the human habit of escaping by automobile to the suburbs. Especially since World War II, urban development is undertaken at high original standards, most notably for streets, utilities, and building codes. But then each area proceeds into a cycle of decline, powered by the escape to newer suburbs. Some commercial strips, even those considered *miracle miles* in the 1950s, had by the 1980s decayed into to drug strips, bypassed by new freeways, and lost to the competition of huge shopping malls. Some bank and supermarket buildings have been constructed and vacated by original owners after hardly a decade of use, having become obsolete in the changing winds of competition. This urban wealth is exploited and consumed as if it were but a line item of corporate cash flow.

When great acreages are made randomly accessible by car and developed for virtually any urban use, a decline of major portions of the city is as certain as the next round of speculative urban developments. That is but a reflection again of the city as being a playfield for developers maneuvering for larger shares of their markets. The city is but another strategic element for their operations, merely another product that is hardly more durable than the automobiles that underwrite the rise and fall of urban areas, section by section.

In this century, great contradictions began to occur when the automobile broke the bounds of urban proximity after 1920. The great urban dance of house and car then began. Developments could go anywhere, and a new freedom seemed to be in the air. But the freedom changed into a build-and-abandon syndrome. Then old unkempt areas of American cities became bankrupt junkyards, structured, it seems, like unprotected food or old clothes. It mattered little whether industry created use-and-cast-away hi-tech products or build-and-abandon sections of the city. The process was the same. In urban terms, that process creates a counterfeit society, good as long as one can get away with it but inevitably false in a time when the whole society must face up to the tragic reality.

The waste is not just wealth or the environment. The lives of people are no less at stake, especially those who cannot perform at par in the exchange system and who must live on the lower slopes of the build-and-abandon process. While the rising standard of living for most people is offset by the multiple inefficiencies of urban life, the poorer population suffers a special offset by every drug peddler imprisoned, unusual health hazards, and every child that learns to live lifelessly or brutishly.

What has been created in cities is a satanic engorgement posing as the American Dream, counterbalanced by a planned obsolescence and a decay of older places for other people who have lost hope of ever seeing the dream. The waste, so systematic and forceful, carries with it enormous consequences beyond it's specific effects. The revolutionary power of the automobile combined with the wealth exhibited by spacious homes on large parcels in the distant suburbs—combined also with the immense expenditures for boulevards and freeways—points to a grave, new power of the industrial and economic development with destruction built in.

Quite plainly, the economic system has become so very powerful that it can support, and indeed promote, a *runaway economy*. The power of the economy is so great that it can underwrite destructive economic growth that in turn demands more economic growth to manage or mollify its own powers, of which once again the enormity of the automobile's impact is the chief example. I have noted herein and elsewhere its vast force as literally the chief destructive force operating in cities. Beginning about 1920 and operating in full force since 1950, the twisted story of corporate strategy and conspiracy, coinciding with the growing wealth and exploding cities, the naked force of corporate automobile urban destruction has become epical. Of course, that force reflects the American pioneering tradition and the rural sentimentality, and also struggle to follow the unfortunate wealthy-class models of mansions and manor houses. But ultimately the automobile's combined power to access enormous acreages and consume much that it accessed was the lever of ecological action that exploited mountains of resources, especially fuels. (It has sent the country into a petroleum deficit diplomacy and military actions in the Middle East, possibly signaling the first global conflict for declining resources.) In demonstrating the industrial system's power and profligacy of the runaway economy, the automobile is the greatest force of gluttony. But others exist, such as the evident in the geometric commercial growth through promotional advertising, raising smaller but significant examples of consumer purchases of other products of questionable sales, including clothes in pursuit of manipulated fashion, tobacco promotions in the face of stark health risks, beer and liquor advertising promoting growth of alcoholic consumption.

However, there is the more tragic side. Through abandonment, the cities' older areas fester into desperation. Exorbitant costs arise from disease, drugs, gangs, and murders. In these wealthy times, imprisonment has never been greater. These are the grisly additions to the runaway economy. Hence, both the leading and lagging sides of society propel the insidious power of the runaway economic phenomenon. Americans intuitively know this but find it difficult to sort out in their isolationist house stocked with television and many little-used skis, boats, campers, or motor homes, and the disciplined distractions of the workplace—all agents of the runaway economy. That economy feeds upon the automobile's insatiable demands, then upon the promotion of distant urban growth. A grand conspiracy could not have done it better. And to a degree, conspiracy lurks behind much of the modern strategy for infinite, naked, destructive growth.

CHAPTER IV

DISTRUST AND DEFENSE

Defensive Privacy

> *Empty affluence, empty idleness, empty excitement, empty sexuality are not the occasional vices or misfortunes of our machine-oriented society but its boasted final products.*
>
> Lewis Mumford

> *The civilizational malaise, in a word, reflects the inability of a civilization directed to material improvement…to satisfy the human spirit.*
>
> Robert Heilbroner

> *Amidst unparalled productive power, freedom is being eroded and confined.*
>
> Paul Goodman

The statement that "we make our environment, and the environment makes us" has become a common aphorism. Yet, since society is passing through history's most radical transformation, it is sad to note how little is known about deep human changes that have occurred in passing through the industrial and urban transformation of most Americans. No simpler example exists, however, than in suburbia, where any plot map or aerial photo reads almost as simply as a social register. The scene from above creates an impression of stall-like cages of a mammoth system of kennels. Unlike a kennel, however, the occupants depart each day and travel great distances—unimaginable a century ago—to obtain the necessities and benefits of life.

We have seen how the fragmentation of land simply isolates human behavior, which in turn requires massive efforts to overcome the distances of isolated living.

We also have seen how zoning and subdivision ordinances separate activities—and the behavior of persons. Zones segregate urban activities by function, followed by subdivision regulations that separate land into blocks and lots for discreet, personally exclusive uses. No *social* goal is sought by either the developer or the buying public. Appeals to social class may be evident in larger lots in exclusive neighborhoods, where occasional country clubs may appear. Churches occur randomly, as does their membership. Schools occur regularly, but hardly break the spread of the private imperative. Shopping centers seek out the boulevard intersections with heaviest traffic. There is no hint of any higher social or cultural purpose structured in the layout.

By design, the American subdivision lacks everything a single family cannot itself purchase and put on one lot. This kennel-like encapsulation organizes behavior for all of the isolation one can afford. Greater wealth simply buys larger lots and greater isolation. Buying a house, paradoxically, therefore means buying into a life with as much solitary confinement as possible. Such barrens offer no commons for casual meeting (for even neighborhood parks are designed to be socially barren), rarely facilities for cooperative activities, no conditions for a social evolution to occur, and no basis for a unity of public interest short of an alarm worthy of Paul Revere. An issue must be painful or threatening to bring people together. Positive social aspirations simply have no basis to appear in today's subdivision design. Stunningly revealed is a *negative ideal of the American city*, an ideal of separateness poured into concrete and hammered into wood, all reinforced by fences and heavy vegetation. Thus made is our modernity.

This isolation extends to individual family members within each house. Separate television programs of special interest to each person in separate rooms are reinforced by individual microwave meals. Thus encapsulated, each family member in each house and each family on each lot creates an overwhelming separation of a person's affairs and psychological disposition. Each activity is conducted in the different rooms afforded by affluence, unlike the past when activities were mostly common.

Today's personal ownership of an urban place puts very special-colored spectacles on every person's view of appropriate behavior, public association, social position, and personal prestige. Emphasis on ownership of separate parcels voids casual interaction, gregariousness, local public events, or discussions of public affairs. Property lines establish stress barriers against friendship and negate a public mind that can visualize higher public purposes.

Consequently, significant *local progress* requires an exponential hardship, almost a false concept. Perhaps the *mine-yours* paranoia becomes a silent censor denying a concept of future social progress by implanting a severe austerity of interpersonal and neighborly behavior. Trust barriers are set high, and a permanent guardedness is sustained. Hence most neighborhoods neither recognize nor lament their loss of companionship. Isolated privacy is the defined goal, set in the physical layout, and social sterility is the expected and self-fulfilling result.

Special burdens are also created when the terms of friendship are reduced to entertaining or being entertained at home, that is, trying to be impressive as a host or being a guest and making judgments about the host's dinner, conversation, or music. One party is initiator and entertainer, the other a captive critic. The atmosphere is structurally unbalanced. There is no neutral ground upon which a truly free evolution of friendship can easily occur. That friendship, if one can overcome the handicap, must be built with an onus upon each party that sadly takes on a character like negotiating a business contract, all with its narrow calculation of benefits and drawbacks.

Privacy in the condition of overwhelming private ownership naturally becomes the defining virtue. Other virtues lacking possessory interest simply atrophy or never appear. Trust, spontaneity, giving of oneself, openness, public values, shared traditions, and public leadership, are consequently diminished. Thus people are channeled by their isolated living to segregate themselves. If we analyzed the psychosomatic conditions of suburban society, the divisions established by property might dominate the analysis.

The divisiveness also breaks into the small nuclear family, where an excessive dependency is created between spouses and with children. When the extended family was formerly unified as both producer and consumer, and events of the family were often shared with a wise grandparent and other close relatives, there could be an easier give-and-take between husband and wife. Neither person had to fill all of the emotional needs of the other nor together that of their children's. But in today's more isolated setting, each spouse of the nuclear family must fill most of the needs and wishes of the other. Under such excessive circumstantial demands, many families that externally appear solid suddenly explode into divorce. The divorce itself also may reflect an unconscious escape from the imprisonment of isolated suburban tracts.

A special privacy also afflicts suburban children raised in the nuclear family. Where population densities are low, most notably in wealthier neighborhoods, there are too few children to find playmates within walking distance. Without casual play during pre-school age, a learning and social deprivation occurs. Parents do chauffeur their children about and do put them into day-care centers. But these steps are burdensome and limited. Whereas large rural families with numerous children once could create their own social milieu, today the single child, or two of distant ages, lives with a social handicap they may carry for life.

The overwhelming suburban privacy imprints itself indelibly onto behavior, which easily becomes *defensive*. The person must defend property and status with fences and locked doors. Here a great torque puts stress on human relations, undermining much free, cooperative, and diversely rewarding social behavior.

Pervasive private ownership throughout the city also works against all forms of common ownership, even against the organizational means for common participation. If one third of the families in an affluent neighborhood have their own swimming pools, they naturally refuse to cooperate in building a common pool. Even a place to build such a pool does not exist. An organization to build and manage a pool is also excluded by the separateness of private ownerships. Thus even where everyone might afford a swimming pool, there are the haves and the have-nots. Sharing is thus excluded by endless suburbia without neighborhood focus, available spaces, or a framework in which to organize.

The very possibility of many kinds of cooperative social action is reduced to hard-core determination of zealots, such as a community theater group struggling to perform a play in ill-fitting temporary quarters by persons who must often commute as much as they participate. Undertaking most community programs must overcome excessive hardships of distances, facilities, and organization. Thus many interests of people simply do no become organized at all.

Please note that I do not question private ownership or its counterpart, privacy. Private ownership is not only valid, it is basic in a free society. I raise the matter because in establishing privacy through a pervasive ownership pattern that separates people, other equal and basic values are virtually excluded in the structural terms of citymaking. Values of privacy and association need not exclude each other, but that is what has happened. That exclusionary outcome amounts to an infringement on human opportunity, thus also on freedom. Unfortunately, privacy for many people apparently has become an idealized way to avoid participating in social affairs, preferring the nonassertive escape of television. When people have never discovered the richness of one another or the human vitality of creating and pursuing goals together, the negative side of privacy inevitably prevails.

Numerous profound obstacles blocking thought about building creative urban environments remain with us. First, living in an era of prodigious product development, we still look to the scoreboard of the standard of living for our basic attitudes about public affairs. At some point, however, product inducements take on the character of cultural bribery. Add pervasive advertising, a taint of a social conspiracy is added. Second, with so few alternatives available today in urban living and in the ways human beings may create together, even a dialogue about present and potential ways of building cities is blocked. Third, the historic frontier still rides high in the pioneering myths permeating our urbanity, with the urban lot standing in for the homestead, mere privacy replacing rugged individualism, and motor homes with names like *wilderness* suggesting the prairie schooner.

Whatever the conditions, we cannot escape the conclusion that cities dominated by lots, houses, and cars do powerfully restrict human thought and action. We have lost freedoms which have not yet figured into our social accounting of human rights. If the immense new range of human freedom afforded by hi-tech industry is to be realized, that freedom lies overwhelmingly in the way we think about cities, build cities—and shape ourselves.

Defensive Society

> We cannot be deceived. Men can and do destroy
> the humanity of other men.
>
> R. D. Laing

> The system...is turning more and more of our
> resources away from the nature of human life and
> into the destruction of it. The justification for
> such procedures [in treating juveniles in the crim
> inal justice system] is that this is the way things
> have to be done in a system of free enterprise. But
> in view of that fact that all the money is wasted,
> as well as the children, this seems hard to
> believe.... We are, practically speaking, uncon
> scious of what is going on.
>
> Joseph P. Lyford

There is another, more sinister dimension of privacy that also is promoted by the urban structure: a growing defensiveness and its self-promoting culture of crime. We note how crime and punishment have grown rather than declined with the expansion of social wealth, and ironically this activity is recession proof. Why should crime grow as a cancer in a society of unprecedented wealth? The question bites deeply, very deeply, in a society of proud democratic traditions.

Any casual inspection of urban neighborhoods will reveal how the city is a very selective filtering mechanism, allocating various degrees of human deprivation and opportunity. Slums and ghettos display not only private poverty but also a public impoverishment: hard street and wall surfaces relieved mainly by trash or dead spaces, few parks and old schools, abandoned tenements as sad as the graffiti sprayed upon them, street-front shops with steel-roll fronts speaking to bitter-end economics, and no driveways or private garages. The automobile, which is the bane of middle-class America, has become a measure of impoverishment for so many people, especially minorities.

These neighborhoods of hopelessness cultivate their own forces for crime in each generation of teenagers. The people who huddle so tenuously to the last places available to them for their fragile security do not know how to escape their devilish inopportunity. Single parents, without helpful education, struggle against odds in the hi-tech world simply to raise their children. Their own background may have been brutal, and this, too, is transmitted in the social genes that each generation passes on to the next. Formal education is seen as dead end, so teachers morally abdicate to the real-world education of the streets. Predictably, gangs build their own ethic of drugs and violence in which they themselves become the most frequent victims and pass on this tragic tradition to the next generation.

Of course, the society at large also suffers, so people arm themselves with handguns, neighborhood patrols, security guards, guard dogs, and electronic alarm systems—all again added to the GDP. We hardly blink at a criminal trial costing the public a half million dollars in prosecution and defense attorneys, judges, court reporters, and court security, often followed by lengthy appeals. Then the felon goes to prison costing $20,000 to $80,000 per bed and another $35,000 or more per year for operation. In monetary terms, the prisoner is king, commanding dollars he could never dream of seeing outside of the reactionary legal system.

While we know that most crime is socially induced, generally from early youth, we still look the other way when the law-and-order society insanely bulldozes a part of humanity into the gaping, killing crevasses of our democracy. Where is the democracy of equal opportunity? Where is the rationality of spending huge tax sums on brutal law and order when a fraction of the cost might have broken the cycles of many criminal careers?

Still, despite awareness of the social sources of crime and knowledge of the costs, the old law-and-order reactions prevail. In the 1980s the national prison population actually doubled to a per capita population ten times the rate of Japan and far above our nearest rival, South Africa.

Let every execution of a criminal citizen slash deeply into our social consciousness, not so much because that level of reaction is cruel and unusual for the crime but because the underlying crime was committed earlier by society itself, reflecting deep, long-developed, socially cruel values. Stopping executions misses the bigger issue. Make no mistake, crime afflicts everyone. Yet the saddest tragedy for society is the unfortunate criminal himself, who but acted out the earlier loss of his self in the crimes he committed. He pays for his crime twice. The first, real tragedy was five or twenty years earlier, when the bestiality of his life's untenable living conditions were internalized within him and turned destructive. The second tragedy of being put to death merely punctuates the deeper crime committed by the whole society.

Have we asked ourselves how many prisoners in our many prisons have in their critical development years ever truly experienced social justice, reasonable personal encouragement, a satisfying home life, a life free of torment, or second chances in a constructive setting?

This blot grinds down upon many citizens in the mistaken view that the society itself lacks the means to create a constructive setting in which to build human value and give positive social direction to the great changes of society. Any comparison of national crime rates clearly highlights the social sources of most crime. What is lacking is inspiration and will. Therefore we react frantically for law and order. Today the youth who are without hope are like Aztec sacrifices to the gods of industrial society. But in our case, they propel the counter-productivity of the crime-and-justice system that accelerates the runaway economy.

Yet, amazingly, even with unprecedented numbers of imprisoned citizens, most people afflicted by the destructive city do not go to prison. Far more people take the afflictions internally while avoiding overt destructiveness. Rather, society's disingenuous treatment of so many citizens breaks out in a variety of other ways: from depressions to psychotic breakdown, from personal coldness to aggressive or shadowy business practice, from exclusivity of social distinction to conspicuous consumption, from life committed to compulsive work habits. Plainly, we are all susceptible, and all of us must respond from within to the very confined social setting of our backgrounds and circumstances.

The first task in building new cities and new environments is to empower the disenfranchised, to offer purpose, real freedom, varied opportunity and—something very much more—a love of democratic society for all of its members. The task requires finding the environmental and social means to restructure the social DNA in our behavior that creates the desperation that penetrates so many of us from generation to generation.

Constitutional *freedoms from* oppressive forces are not enough. Freedom will always be threatened until we build a *freedom for* better living for everyone. And that newer, wider freedom shall rest on positive social and economic opportunities, not merely legal and political protections and reactions.

Built For Disorder

> *There is something self-defeating about our tech-*
> *nological dynamism: it is unstable almost by defi-*
> *nition, because those who further it believe in*
> *instability, and do not realize that without conti-*
> *nuity nothing that can be called progress is possi-*
> *ble.*
>
> Lewis Mumford

After nearly two centuries of building cities in pursuit of wealth, Americans feel a creeping defeat stalking not only their progress, but also their safety, even their sanity. Widespread anxiety grows steadily toward a sullen but desperate public paranoia. And so today, security permeates behavior more than ever. Yet there is really no visible prospect for significant and general improvement. Americans feel overwhelmingly trapped by cities they have built, at best something that must be tolerated, at worst something that threatens life, limb, and property. This seems to be all but inevitable in the absence of a higher goal for cities.

Our cities were developed for business profit and to produce profusely. This they have done. But escape from the cities to something better became deeply a part of the process. Pressing for more law and order is another grasping for security, following the mass escape to suburbia. Escape has been available for those who can afford the move, while oppressive law and order has to suffice for those who cannot. Yet escape to suburbia as well as law and order both signal a conceptual dead end for cities built by and for economic motives over the last century. No higher vision was called for over the decades. And so none is forthcoming now. Thus we must ask where our idealism has led us. And where is it now leading us?

The very small number of public-spirited people who work for new and larger museums, art galleries, parks, and zoos, or for downtown revitalization commendably demonstrate special visions—of great worth, to be sure—but not a *comprehensive urban ideal*. They are heirs of the people of 1900 who had a vision of abundance, a kind of a vision that might have been called great, even if they could not have predicted the physical, ecological, social, and even economic disorder that affluence would bring. Somehow sights have narrowed in the intervening decades. Pathetic it seems, then, that visions of life in cities may have narrowed precisely when the human sights could have been justifiably raised and new grand goals might have creatively motivated people, at least in the last half of the twentieth century.

Cities built for escape and abandonment cannot avoid disorder, that is, the deep disorder the whole society faces today. Yet people seem to shrink from even daydreaming about what cities might truly become. They shrink away, even as the immediate options of urban life are narrowing, even as the urban quality of life erodes, or even as the simple bottom line in the standard of living no longer reflects the evident numbers.

And the disorder propels itself. With money, people escape not only to the suburbs but also to second homes along shores and in mountains. While they describe their escape in beneficial terms, the mood is becoming increasingly negative there too. And quite so, for the people build the same urban problems in their recreation areas as they do at home.

The very mood of escape itself has its own disturbing face, quite aside from the urban debris it generates. Escape to outer suburbia means no commitment, no responsibility for the common good. The citizens who escape the city also escape the responsibilities of citizenship. We may cry for democracy, but escape does cannot serve freedom.

We live in a time of the most exacting and comprehensive interdependence of people in history. Yet, fracturing the urban commonwealth is sadly accepted as inevitable. Thus urban escape and the runaway economy become nearly synonymous. The economy grows upon its own chaos. While that process may be self-integrating in purely economic terms, the vital unity of city and citizen is destroyed by the onrush. Economic integration and monetary progress thrives on social and urban disintegration, as it does upon genuine needs and desires.

Escapes to the suburbs and resort areas build on the continuum of increasing social isolation of the family. In urban terms, there is no resting place for the body or spirit. The spirit is indeed lonely today, since it must face up to the subversive ideal of family isolation and the forces promoting public dissociation and dysfunction. Television, itself addictively isolating, wins the day, as does compulsive buying. But stresses of personality result in a boom for clinical psychology. Yet, even the happiest working through of a psychological disorder with a counselor merely returns the patient to the same threatening conditions that are as cold and threatening as ever. Then it is all too easy to withdraw into more reactive and forbidding modes of behavior. Trapped, we have thus become as confined in old visions as much as physical forms. The challenge for our time is to break up this dooming vice.

Structured Barriers to Friendship

> *It is this very American protection of individual-*
> *ism to the exclusion of cooperation and commu-*
> *nity that may be in large measure responsible for*
> *many of the psychological crises in our society*
> *today.*
>
> Rosabeth Moss Kanter

In the anonymous and cosmopolitan world, Americans do not know how to structure society to make friends. The many barriers erected against comforting association reflect the city made by social atomization. By their very terms, the privacy and isolation of houses and apartments, cars or mass transit, create settings nearly forbidding to pleasant, casual encounters or heartwarming interactions, let alone continuing relationships.

Time for friends becomes an issue when commuter hours increase and monopolize one's time. Active memberships in metropolitan museums, theaters, and arts societies are discouraged by the shortage of time and the simple burden of getting to them. Building solid friendships usually requires free and casual interaction among many persons, that is, a congenial place and ample time for the chemistry of people to freely *test* and gradually develop confident relationships. But there are scant places and little time for such a natural growth to happen.

We have today a society starved for friendship in the atomized city. Some persons consequently seek out association by compulsive, grasping efforts. This means all-or-nothing overtures by one party, and the very circumstance virtually dooms the try. Complete withdrawal is then common, and greater self-inflicted isolation is the result.

With few places and little time for friendships to grow naturally, the gulf has widened between the sexes. Casual and natural growth of association is critical, especially in building first trust. Unfortunately, considering the anonymous character of today's cities, we also teach women to be distant and defensive, sometimes, unfortunately, for good reason. Similarly, men are expected to exert macho or money values, perhaps also with misleading results. Therefore, every expression of openness is all too easily interpreted as a sexual advance, a premature step of serious courtship—or an effort to disrupt a marriage. Indeed, the best kind of courtship may occur when founded upon a broad base of casual associations in which trust can grow prior to courtship and both persons can know each other beforehand in a variety of circumstances.

Indeed, such a trust built within a group may strengthen an existing marriage. Regular, casual association in pursuit of hobby or arts interests by one spouse may relieve the monolithic interdependence that both spouses so often put upon each other when their marriage is isolated in a tract-house setting.

We can imagine social and urban conditions in which varied relationships between the sexes can be non-threatening, free, happy, and fulfilling. Fundamental is a kind of community (itself a trust setting) in which there is a close interaction among a limited population with many social options close to home. These conditions exist today only in fragments.

Varied human friendships are basic to a prosperous personality. In the modern condition, where the only values and qualities of life that prosper are those consciously pursued, our tragedy is that the most personal and familiar values, if not the most intimate, are those that not pursued and thus suffer most. Poignantly, society does not organize to build congenial support for human relationships. Human association, which is the heart of human prosperity, suffers and so often tragically passes away.

We need not praise traditional values as ideal, for we know that historic human relations suffered many brutalities and systematic deprivations that we have now happily cast aside. It is for our age to recognize our own systematic deprivations buried in today's structure of life. Yet it is more important to recognize the enormous human possibilities. Such a vision need not entail stiff social strictures or an imposed moral righteousness, as some traditions have demanded. We do need to search for a new framework of ideals, for their best fulfillment will necessarily be found in the city, which structures our very humanity. When we do that, our search for solutions to grave social problems, and for simply better friendship among ourselves, will come into clearer focus.

CHAPTER V

URBAN DISCONTENTS

Radical Ecology

Development of the modern American city is a history of human revolution, despite the absence of the historic drama of other American histories like industry and science. The city frames and shapes the results of technology, even defines what technologies are appropriate, or is misshapen by them. *American urbanization is perhaps the concluding chapter, and also a part of colonization, westward movement, and industrialization; in human terms, it is at least as fundamental as technical advancement and the outpouring of new products.*

Never a basic aim of the greatest transformation in history, cities impacts upon life are massive, yet seen as incidental outcomes of economic expansion. While production is mainly a technical and economic phenomenon, and changes readily, cities are overwhelmingly a human phenomenon and are, for most practical purposes, permanent. While historians have long acknowledged cities to be central to civilization; indeed, we could say, *cities are civilization,* as their common root indicates. They are treated as sterile and mute artifacts of the great changes taking place; and are seen as inert containers, not as dynamic shapers and movers of civilized life, only reflecting the true "operative" realms of society.

So, tragically, today's cities are ill formed, bloated, enormously inefficient, and socially regressive, and are now setting society on a course toward ecological disaster. Essentially, they have become mechanisms merely to promote economic growth; consequently, their social significance is understood to center on private wealth expended privately. Thus varieties of social development are excluded from current economically determined urban form.

Behind it all, there is a lack of philosophic, scientific, and idealistic focuses upon the city. Lost is a basic and unified urban theme in society, which highlights a prophetic omen for the yet unrecognized threatened defeat of civilization by their cities. That is especially acute because cities should be the instrument par excellence to advance social and ecological development. In the United States especially there is an inordinate dependence upon machines and techniques, a dependence which masks important underlying promoted spatial, material, and human inefficiencies, that is, encumbrances embedded in the urban physical structure. This great waste not only undermines a sustainable human ecology; it also shapes an unheralded human suffering from the profusion and disarray of goods in ill-formed, anti-social urbanism. The lack of public, coherent, and critical concern for the city, despite today's urban planning, reflects the view that the city is mainly a public arena for expressions of private values without a reasonable counterbalance of public values with clear measures of social well-being. *Indeed, the telling and prophetic economic fact is that the more wasteful and destructive the city becomes the more it contributes to the GDP.*

The ironic philosophic vacuum surrounding the city, despite a stacked pallet of urban books, underscores the serious lack of a fundamental agreement upon ideals of urban form and civic excellence, reflecting all too much the silent doctrine that chaos must prevail before even minimal reform will be undertaken. There is good reason, therefore, to say that the city is virtually a non-entity, almost a non-fact, and hardly a manifestation of the "real" matters of society. Also prophetic perhaps is the apparent fact that the first word to Americanize the English language involved drawing "lots" while still aboard the Mayflower to determine the allocation of parcels in the new town of Plymouth. In any case, American citymaking always involved a high degree of chance of the marketplace, which in turn emphasized division, separation, and social fragmentation over integration and unity of urban form.

Endless Expansion

Coming to a new land forced settlers to make conscious decisions to form their new towns, having removed themselves psychologically as well as physically from their old world environments. On new land most long-held traditions of properties and buildings were bent to suit the new circumstances. And on a new continent a sweepingly simple arrangement was called for. That arrangement was the gridiron plan, highly rational, universally applicable, and endlessly expandable; it could accommodate most activities admissible to the city. Easily laid out, it could be marketed by mail, and was manageable without much management at all. The straight streets and rectangular blocks required no agonizing decisions. As a starting point, a road intersection in the wilderness would do. The grid could be applied to an existing informal layout, as in New York when the gridiron was adopted in 1811 to rationalize the City's northward expansion.

The rate of urbanization was itself revolutionary, starting in 1800 with but 322,000, urban population grew to 6,217,000 in 1860, to 54,158,000 in 1920, and 149,325,000 in 1970. This represented a growth of the urban population from six to seventy-three percent. It did not seem to matter that this growth was reflected in land values that always pressed against the worker's ability to pay, and therefore that tenements built in New York after 1835 often covered ninety percent of each parcel and contained many rooms without light or air, or that health conditions constantly threatened epidemics. But economic arguments could not be contested.

The fifty years following the Civil War may well be known as the most dramatic and pivotal in American urban history. If the railroads completed four transcontinental lines by 1890 and created a nearly unified national production and marketing system, the new major urban centers spread their domination to ever greater regions, and became instruments of the transition no less than the railroads and industry. The multiplying industrial output demanded a concentration of population no less than new industrial processes, along with the railroad's own demand for steel rails and rolling stock, the increasing and more varied farm machinery, or the growing demand for factory-made consumer goods. The only medium capable of serving that array of development was the ever-growing city.

The industrial city that emerged around 1900 was like no other in history. Industrial tycoons replaced the landed aristocracy in social leadership. Large factories replaced home artisans. Markets no longer characterized teaming town plazas and took on the forms of grain and stock exchanges, department stores, corporate offices, and huge stockyards. And the city of 1920 brought the country to the first demonstration of mass consumerism, including the largest consumable in history, the automobile.

With a new scale of population, longer urban distances appeared, and formal methods of transportation were required. Horse-car trolley lines appeared and by 1890 claimed over 4000 urban miles of track nationwide. Costly subways were introduced in 1897 in Boston and grew to a few hundred miles by 1920 when their extensions came to a virtual halt (until late in the century and a few new subways systems appeared). The automobile apparently halted the first round of growth but its congestion stimulated new systems late in the twentieth century. There, too, the automobile promoted the development of new transit systems because of its habit of congesting nearly every new freeway. Especially the new subway systems became safety valves to relieve pressure on overcrowded roadways, most notably in Washington, D.C. and the San Francisco Bay Area.

Powerful changes in urban life also appeared with the advent of electricity and telephones. New York recorded the first steam-generated electric utility plant in 1882, built by Thomas Edison, primarily to utilize his own incandescent lights. Telephone lines, first strung in the late 1870s, served some 250,000 locations in 1890, and 2,000,000 by 1900.

The American city of 1890 to about 1910 almost seemed to "invent" success. Until then the citizens were proud of the vast growing size and took pride in ever larger cities, before a patchwork of new cities on the metropolitan fringes reflected a rather sudden rebellion against such massive cities.

Henry Ford could never have built his cars without the vast resources available in Detroit and other cities. Cattle barons depended no less on Kansas City and Chicago, wheat farmers on Minneapolis. Silver was discovered on the barren slops of Virginia City, Nevada, but its developers quickly brought their wealth back to San Francisco, 200 miles away.

Sacrifice of Cities

Success and growth have become united and supreme in both the business and public mind. Problems along the way, it was thought, could be handled with more technical knowledge and growth of wealth. The American Dream was simply onward and upward. But cities themselves were *not* part of the Dream, however essential they were to every part of it. What mattered was success and wealth and, when translated to economic growth, the wealth was cemented in the mind of business and government as the foundation assumption of ever growing wealth. All spheres of human progress could be subsumed under produced wealth measured as infinite economic growth.

That ideological assumption remains largely unexamined to this day and stands behind the irrational growth of the runaway economy and today's twin social and ecological disasters-in-the-making. What was sound growth up to the achievement of good living was not sound after affluence dominated—which introduces wild factors into the seemingly smooth running economics. When wealth reaches mischievous levels of supposed freedom of the consumer, in reality it underwrites vast new exploitive marketing possibilities for producers. Since cities had no independent ideal for their service to society, they could not be part of the measure of human progress. So they were judged mostly by their faults. But they were excellent in inducing a waste of wealth. And the faults that counted after 1920—roughly when the problems of cities began their massive transformation away from the faults of poverty to faults stemming from affluence—centered on making a place for the force of the automobile, its congestion in traffic and lack of parking.

Cities could decline as effective environments, but the big money went to roadways, first boulevards in the 1920s, then into the gold-plated pavements of freeways after 1945. They cut through old neighborhoods and their noisy environments contributed to urban decline while they made auto commuting to the suburbs feasible and sprawl attractive with new personal wealth. Thus the wealth arising from economic growth fueled both the escape and the urban decline in the old central cities, sacrificing them to spurious progress.

The base assumption, that the growth of productive wealth is universally beneficial when it is applied through the market forces of real estate formation of cities, aided by politically forced public construction of roadways, illustrates how the absence of specific human goals, especially the lack of urban goals, can have powerfully negative results, both socially and ecologically. Once the ironclad necessities of food, clothing, and dwelling could be met, then subsequent growth of wealth too easily promotes serious public issues. If wealth is a power and if the legitimacy of power is always questionable, no more poignant case is evident than the wealth applied to the dysfunctions of cities. Cities were sacrificed to auto transportation.

And no better case is needed to demonstrate how economic growth is only an economic goal, not a human goal, and is in reality, therefore, essentially a non-goal, not valid when society shifts from scarcity to abundance and its urban, social, and ecological consequences. Economic growth thus highlights how wealth, and especially the self-propelling economic growth, endangers human values and destroys human purposes. To encourage economic growth has become the ever-present priority of public affairs. It rules over other social matters and demonstrates how powerfully economic growth has become the central issue in an advanced society. The case most demonstrating that torque of human affairs is the city.

Sub-Urbanizing City Life

If after 1920 American cities began to show the first signs of distorted progress, it was because of another unprecedented development: the sub-urbanization of the metropolis. The term sub-urb or suburban is very apt, indicating something less than completely urban. The earliest official use of the word "suburb" is associated with the 1880 census, which coincidentally referred to the word "metropolitan," which then applied to only New York City.

Four factors generally account for suburbia as we know it. The first was based on the system of land tenure and the nature of private real estate development that promotes random, market-oriented, hop-and-skip growth on any available land. Second, a persistent rural image was maintained as an ideal based on now misapplied rural values. Third, the low densities, whether in lots of 6000 square feet, a quarter acre, or the prestigious acre itself depended importantly upon the motorcars' ability to access huge quantities of land in much of the rural area around cities. And it is the car operating in conjunction with the one-family, detached dwelling that lies at the crux of American metropolitan development today. The fourth factor was the powerful motive to escape from the old decaying inner city, partly because of the multi-obsolescence of older dwellings and partly the affluent ability to afford expensive new dwellings.

Most old houses became obsolete because of the costs of retrofitting them for heating, cooling, lighting, bathrooms, kitchens and their new appliances. Remodeling was discouraged by the low resale potential, itself a reflection of the escape mentality. These biases were reinforced by the many building code and banking hurdles, as well as ethnic prejudice and social attitudes favoring single-family houses. At first the escape seemed to be largely aimed at the unpleasant remains of the nineteenth and early twentieth century city. But after 1945 the growing metropolitan populations with greater affluence and the ethic of increasing consumption combined to expand the escape. Thus when the middle class moved to suburbia, "one could escape the urban conditions—crime, grime, congestion, poverty, high taxes and bad services, inadequate schools, pollution, deterioration," wrote the student of suburbia, Louis Masotti.

The great post-war flight could not have taken place without the growing wealth reinforced by seemingly unlimited land, energy, mobility, productivity, and the pressure of growing urban populations and the increasing class exclusions. These were precisely the conditions compelling the accelerating industrial demands upon resources that opened a new radical chapter in human ecology. Whatever the novel experiences that expanded consumerism conveyed, the growing production and consumption embedded in the structure of the city set society on a collision course with both human values and the limits of nature.

Suburbia greatly modified the classic American gridiron pattern. The new profession of traffic engineering was established to improve the more important part of the pattern, the major roadways of heavy movement. Where the original grid established an equality of access between streets and properties, with easy exchange between them, the ascendance of mobility meant that access would serve traffic on expanding roadways over the uses of land—thus the term "limited access highways." Urban development itself got little innovation, skill, or assistance. But subsidies for movement reached to the limits of private wealth and public budgets.

According to the massive new functionalism, two new levels of the "gridiron" appeared. Some streets after 1920 were widened to become major boulevards with four, six, and sometimes eight lanes. These were usually on section or mid-section lines, which formed huge, super blocks of 160 acres over the old blocks of 3–5 acres. Then freeways began to evolve in the late 1930s, requiring wide swaths of land for four or more lanes, wide medians or barriers between opposing traffic lanes, no direct access from adjoining properties, and interchanges with boulevards. These freeways essentially formed new grid patterns of 5–10 miles. While both boulevards and freeways were responses to increased congestion, they also greatly encouraged the extension of urban sprawl into the far countryside. Then, access control completely favored movement over an access integrity of local uses of land.

The new specialized roadways also coincided with the major reaction against the old blighted and mixed industrial, commercial, and residential activities and introduced specialized uses of land segregated by zoning. The operative terms there are reaction and segregated, for reaction did not represent a new urban ideal and segregation helped to increase the distances of commuting, thus propelling the demand for boulevards and freeways. As one planner noted, "instead of guiding private development," the planning "merely sanctioned it."

The consequences of sprawl, land segregation, and the great spaces and their great distances demand major trafficways, the individual was forced into a diffused, wearisome engagement with entire cities or great parts of metropolitan areas to become a minimally viable citizen. This meant that the less integrated and more dispersed the city became, the burden of "integration" fell upon consumptive mechanical transportation, which then promoted more dispersion. Then public policy sought to extend transportation as an endless good, always to be expanded. Hence the suburb seems to have overturned the city with its new chief characteristics of sprawl, segregation, and masses of moving vehicles. It made the city barely viable and socially and ecologically recessive; that is, it created a massive, mechanized, irrational dynamism. *Thus problems of 1900 were mostly problems of poverty; but by 1950 the chief urban problems arose mostly from wealth, an ominous outlook for both people and nature.*

Giant Functionalism

If the American city grew rapidly with few constraints of history or ecology, its nature came to closely reflect the forces of industry and commerce under the aggressive corporations. They enjoyed investment initiatives and generated the new traditions of promoted consumption based on new wealth. Entrepreneurs could claim that the rights of private ownership and free enterprise were the centerpieces, not just of economics, but also of the entire society. This not only depressed public leadership but also dominated public budgets for the roadways, required by mass automobility, while denying a balanced funding with many other public facilities and programs. The effect was to keep public eyes on the endless need for new roadways required by automobility, importantly as a foundation for endless economic growth.

As families became less self-sufficient in the city for food, clothing, and dwellings, they became more dependent upon jobs and wages. *Then, upon the rise of widespread affluence, the conditions of money dependence soon created the power of economic determination.* Money could then dominate private values and public policy as never before, as recessions could give the country aggravated paranoia reflecting the trauma of the Great Depression. Money became all important, more deeply affecting career decisions, and its purchasing power became central in acquiring new houses, appliances, cars, and recreational equipment. Simultaneously corporate powers grew upon product strategies, nationwide marketing, and intensive advertising. Economic determination ruled ever more powerfully as money could purchase all of the many necessary things of life and many other benefits like cruises or fast trips to Europe. But a bondage came with the new pleasures, the economic determination that narrowed more strictly upon production and consumption, good certainly for the first generation to experience wealth, but hardly a basis for a subsequent lifelong philosophy of expanding meaningful human experience.

As mass consumerism got into stride in the 1920s and with new force in the 1950s the economy itself passed through a transition that could be called revolutionary. Whereas primary initiative had long focused on increased productivity to promote economic growth, the appearance of general affluence increasingly shifted attention to marketing as the primary means to promote growth. This transition shifted marketing toward luxury goods for the home, recreation equipment, television, and travel. Since considerable wealth went into plumbing, electricity, air conditioning, and costly appliances, dwelling obsolescence accelerated, and these events were then also associated with larger houses, additional cars, and sprawl-generated commuting. The prize was greater comfort and more diverse activities. Also powerfully evident were the new serious downsides, the defects of sprawled locations: longer distance commuting, greater traffic congestion, more functions (to service the house, yard, and cars), and family isolation.

The shift in emphasis from industry to commerce applied to expanding the economy through increased consumer demand centered importantly upon one product, the automobile. No other product, not even building farm-like houses in the city, so powerfully transformed was the city's form, function, and meaning. And it remade the city mainly in the short period from 1920 to 1970, especially after deducting the depression and wars. It pushed the city into becoming a built-in escalator of consumption. Annual auto production leaped from 4192 cars in 1900 to 181,000 in 1910, and then soared more than tenfold to 1,905,000 in 1920 when its serious impact upon urban form began. Both automakers and motorists bitterly complained about the city's inability to accommodate traffic and parking needs, no less than they had earlier languished in rural mud. And little wonder, for what they were asking for and got in only five decades was nothing less than a complete remaking of the urban world, a giant new scale of urban automobility as part of the largest manufacturing and construction enterprises in history.

A first watershed of automobile development occurred in the 1920s with a regular system of auto purchase loans, an organized used-car market, general production of enclosed bodies, and styling as regular elements designed to promote sales—from "a mass market to a mass-class market," said General Motors president, Alfred Sloan. No less fundamental were the changes to make the environment suitable for distant travel. Based on the Federal Road Act of 1916 financing underwrote the rapid growth of an unprecedented road construction after 1920, enough to build a national network of paved highways by 1940, and a new scale of political power surrounding automobilia. In the 1920s appeared the first off-street parking and early parking garages. Parking meters were slower, not appearing until 1934. Of course, the service station, repair garage, auto dealers, and wrecking yards became universal in those two decades. Hardly surprising, nearly every major urban innovation appearing between the wars was in response to the automobile: the road house or tavern, motels, varied drive-ins, the supermarket, discount department stores, the shopping center, the industrial district, and an innovative form of urban blight, the great American commercial strips along the widening boulevards.

The radical achievement for the automobile was the freeway, the most rigorous and costly instrument to serve the most costly and least efficient form of urban travel. Sometimes billed as the only motorway equal to the performance of the car, freeways generally serve well until a saturation of traffic stalls them in stop-and-go congestion, when their capacity dramatically declines. So today the traffic engineers are desperate to maintain capacity, introducing diamond, multi-passenger lanes, and signals to moderate traffic entering freeways, reversible traffic lanes, radio and TV traffic reports, and TV monitoring of great interchanges.

Freeways became revolutionary in another way as well: cutting through the raw flesh of the city. New York's greatest freeway builder, Robert Moses, aptly stated the case: "When you operate in an overbuilt metropolis you have to hack your way with a meat ax." ("New Highways For a Better New York," *New York Times Magazine*, November 11, 1945) Put another way, to achieve the integrity of automotive movement (especially through controlled access), one destroys the integrity of human environments. While urban integrity and unity has rarely been applied to urban form, a young black person living in a deteriorated environment can hardly miss the point that mobility for the affluent counts, people in neighborhoods do not.

What is observable to anyone is how urban automobility *disintegrates* urban form, and in addition to its own integrity of movement, demands more movement, more feeder boulevards, more parking, more escape from auto-impacted parts of the city, longer utility lines, more school busing, and more police and fire protection, altogether ballooning consumption, waste, and human disarray. Zoning fits the expansion, organizing functions into neat classes of isolated behavior, based as it is on an essentially negative vision of the city, that is, of zoning unsightly activities out of sight. The misconception, however, was powerful enough to become the principal foundation for "comprehensive" planning after the surging decade of the 1920s. When zoning is set alongside large home lots, the distance-spanning automobile, and space-consuming roadways, the critical components of giant urban functionalism were put together to generate an inhuman urban disintegration. While people frequently stress the wonders of mass production, they blind themselves to a higher social potential of the environments they live in and then must suffer. Since people have struggled many generations to achieve the wonders of new products and great output, it is extremely difficult to understand the self-defeating conflicts created by output that invades the city, but is passed off as the price for progress. But that is what we must expect when we have built our cities as mere corollaries of industry.

The lack of an historic and philosophic guidance of the city as a basic and unified theme is prophetic to the still unrecognized defeat of the city as a valid social, ecological, and indeed, economic instrument of society. What we now see is a new, inordinate dependence upon machines, a dependence masking massive spatial and material inefficiencies, and the inner bleeding of humanity rationalized away as wealth. *Unfortunately the city has become a public arena mostly for expression of private values and corporate power without a counterbalance of positive public values, and these are nakedly expressed as production and consumption.* This system of corruption must be suffered grievously, it seems, before society awakens and takes even minimal action to resolve important issues that don't happen to be within the classic line of progress. That is our conditioning of mind and why there is good reason to say that the city is historically seen as a non-entity and but a shadowy reflection of the "real" facts of life.

It is no insult to technical progress to say that society has applied its tools of progress badly. *If industrial development undermined the vitality of historic urban form, we are dependent upon industry to sustain the exploded cities, their vast waste making, and their utter defeats.* Our inventiveness centered on the parts and pieces of technology and leaves out the creative potential of the city as a unified system with an integrated form built precisely around *human needs*. Society, rather than turning its inventiveness to improving the living patterns in the city, and exploring the varied opportunities of community, or consciously developing a cosmopolitan magnificence, has quite literally turned its back upon the city and built housing in neat but devastating isolation tracts. *Most people thought they were achieving the best of both town and country when in reality they were destroying both.*

It is an astounding ecological fact that Americans prefer the same kind of transportation and same kind of dwelling in both rural and urban areas, despite the enormous differences between town and country. As yet, not even the environmental or energy crises have taught them that there are massive and fundamental contradictions within their individually sought pastoral values, their strident technologies, and the way they think about urban living. *If the city is by necessity a quantity, quality, variety, and intensity of services and activities available to all, then any actions that diminish these conditions also diminish the viability and benefits of the city itself.* This, I contend, is close to the central issue of urban development today

Civic Depression

The American city, with accelerated growth after 1865, may be characterized two ways. First, its population, productivity, technological improvement, evident affluence, and its remaking for industry, air travel, and especially the automobile reveal an unprecedented dynamism. Each fifteen or twenty years of automobile development after 1900 inaugurated a radical transformation to support automobility. Second, slowly but undeniably, a creeping sense of inner defeat has grown from the same soil in the same periods as the successes, that is, from wealth. This defeat, as broad and penetrating as it is, is vexing. And so the critical questions and accompanying malaise are far wider and deeper than any specific issues like congestion, urban decay, or crime. Profound questions are required, therefore, to get to the more critical urban issues. We also need to separate today's questions arising from affluence from those that existed in the ages of unrelenting scarcity, for an acute inadequacy of property then defined the basic issues. But wealth now defines them, despite large populations remaining in poverty.

One should not be accused of praising the American urban environment of 1900 to say that it was vigorous and materially progressive. There was, to be sure, consternation for unrelenting slum conditions. But no one then needed to worry about the viability of the urban centers, let alone sprawl. As wealth increased and backbreaking drudgery was reduced, the harsh living conditions of most people were being slowly ameliorated. Yet the great hopes for basic urban improvements were directed to relieve the growing traffic congestion and to such projects as clean water, sewers, school construction, fire protection. But almost completely absent were projects that consciously sought more positive, if not creative, living patterns, and a social vitality, not even in some large housing projects that might have been understood as ongoing experiments. Instead the projects were seen simply as welfare and therefore deserved minimal thought and stringency budgets. So, after the depression and wars, attention turned solidly upon economic growth, and centrally upon autos, freeways, and sprawl.

But in the period after World War II the wealth-making formula began to demonstrate grave weaknesses. Wilfred Own trenchantly commented, "the American city lacks everything from which no profit is possible." And the problem is more profound than what people might miss in their daily lives, argued Owen. "When anyone demands public controls on the pursuit of the dollar, we cry that our freedom is in danger. But we use the freedom only to despoil and destroy." Of course, the attraction of campers, boats, skis, two cars, and a suburban split-level house were powerful attractions and in the end favored the freedom to despoil. Television, sound systems, computers, and the Internet relieved the suburban isolation. These made it easier to ignore the potential values of a higher urbanity, which most people could not know or conceive. Ominous but deceptive, however, was that a clean white wood-frame houses with two cars in the driveway itself does not seem to signal the escape psychology or the growing environmental deficit of the middle class it represents. Nor does it clarify its antithesis: the inward deterioration in large older sections of the metropolis or declining energy resources. The quiet, tree-lined residential street hardly reveals the pattern that causes Americans to use energy as if it were a smog that would dissipate in a day or two. But behind the pleasant facade appeared the nagging, burdensome feeling that all was not right in the American idyll.

By the late 1960s, freeway revolts became more common and some shifted from fights over location to resounding rejections of the freeways themselves. San Francisco forced the state highway department to dismantle the elevated Embarcadero freeway. But solutions in the end were traditional, pragmatic, specialized, and responses to narrow if not special interests as before. And the issues did not get to the deeper crises, let alone reveal higher urban values, or respond in a positive way to the riots and bombings of that era, or to why the hippies took leave of virtually the whole cultural heritage.

The irony of destructive urban form is that the pragmatic direct-response solutions in the end created the giant runaway economy that is precisely the modern problem. Not only is waste the issue. The total effect is utterly inhuman since it reflects the absence of a strong social ideal, the lack of principles of urban and universal efficiency, a basic structure to shape ecological sustainability, and an ethical foundation for the organization of society. It did, of course, reflect the limited purposes and overwhelming powers of corporate economics.

While the city expanded as a classic market process for short-term profits, the resulting urban form and the draw upon resources exposed the most permanent outcomes of modern urban development. The city is built as if it were a consumable product, while in fact it shapes people's lives and profoundly organizes society, sometimes for hundreds of years. The finite supply of land and the natural habitats are managed in the marketplace to be "consumed," that is, to "waste and squander" in the old definition of consume.

Since we organize society mostly to produce and consume, we have lost critical insight of what an urban, democratic tradition could be. We simply substitute it with mass consumerism. We prevent the development of stable human organizations and environments that might prompt the growth of higher human values. That, I argue, is the eminent role for conscious development of urban community, an organization built around the person in vital association. But today economics rules, especially when production exceeds basic needs and converts excess output to the immense urban wastes of the runaway economy. Then human purposes in cities are subordinated to economic promotions and profits.

Cities and Environments

Only vaguely recognized is the fact that an urban revolution was necessary for the Industrial Revolution. Yet in human terms it was as fundamental as industrial work and the new products. But unlike industry, cities were never the aim of the greatest transformation in history. They were but a gigantic sideshow, treated as an incidental effect of economic expansion. *While production today is mainly a technical, economic, and temporary phenomenon, cities are overwhelmingly a human and permanent phenomenon.* Yet cities grew mainly in accordance with economic conditions, not social aspirations.

Although cities are the most definitive and vital human environments outside families, vital for all qualities of living, their form is governed by the real estate market, which guarantees that their "form" is governed solely by the smallest bits of exchangeable properties. This (a) guarantees that there would be no organized social relevance of urban form greater than the single-family dwelling and (b) leaves the form and scale of the city open to the requirements of the greatest consumable product, the automobile. The family lots formed into blocks and fronted by streets put profits over people, movement over place, and when the pattern is expanded to sprawl, permanent congestion over true spaciousness.

Since cities grew largely in response to economic conditions, their structure was strictly determined by people's ability to pay. Then with the increase of wealth and the arrival of automobility, cities themselves became the primary commercial forces promoting economic growth, partly replacing industry as the leading force behind that growth. In the earlier period of industrialization, the large, older cities were often dominated by the construction of densely-built, low-cost four- and five-story walkup tenements covering virtually all available land—and this gave an undeserved bad rap to all high-density living. But after productivity expanded, higher wages increased the market for consumer goods and the spread city of large lots, multi-car ownership, and the universal pattern of boulevards and freeways. This meant that congestion arose from mobility, not from high density building, and arose especially from transport-demanding urban sprawl. The second and third family automobiles, large lot and house, the massive boulevards, freeways, and parking structures completed the automotive conquest of cities. It was a city built at the outlandish scale of the automobile.

Altogether, a uniquely modern self-generating economy dominated society. Distant low-density development promoted vehicular congestion, and the struggle to relieve congestion demanded immense private and public outlays which promoted growth of an ever-greater GDP, thus resulting in an economic imperium founded at the expense of cities without its essential human rationality. But lost living was more than direct losses from the car; it was also embedded in the isolated, car-only accessed suburban house, an isolation relieved only by the timely appearance of television in the 1950s, along with accumulations of skis, boats, dune buggies, campers, and travel abroad.

The mobilized American city was a sacrifice to economics, which we label the standard of living. The city was an economic stride, filled with a growing range of consumer choices, although the choices were mostly limited to purchasable things or services. There was little that could be said about social advancement beyond the procured benefits, how people might relate to each other with a grand freedom of ease, or how cooperative endeavors among people might be easily facilitated. Lost was a true urbanity of relaxed association with casual and regular involvements at theaters, galleries, plazas, and sidewalk cafes. The car that was necessary to get to one of these places made every association the burden of an appointment and trip plan, hardly a way to enjoy urbanity in a relaxed cosmopolitan setting, especially when accessed in the mood of parking in a six-level garage.

Plainly, an empty consumer lifestyle, possibly a defeat of urban living, certainly a series of social crises, is present in our contemporary urban settings of vast subdivisions scattered throughout large urban regions. But that is our legacy of industry and productive wealth. Unfortunately what critical issues are dealt with are likely to be considered in isolation from all others, not to mention the great range of positive social potential. As essential as the specialized services might be, whether police, welfare, health, or education, they are not generally connected to each other; and they do not seek social vision; or, if they did, there is no public means for them to act creatively. No high urban goals are to be sought, no visions of the good life linking people together outside of social class pecking orders, that is, no concept of what society should do or be for people *as people.* There is no vision of how society should organize and assist people in forming distinct beneficial lifestyles. The abundance of product and service hides the fact that there are so few positive social opportunities to associate and build *personal* ambitions comparable to the pursuit of professional careers.

Hence, most of us live within shells of isolation and alienation, an anomie and harsh emptiness that grips nearly everyone, an emptiness felt most deeply after the loss of a loved one, a sudden divorce, or the loss of a long-held professional position. Each of these events today casts upon us an unwholesome sea of unhelpful faces, unsupported by the no longer existing extended family, or by friendly faces who would at least understand and might assist us. The alienation penetrates us, leaving us with a subversive neuroses that might show or not, and usually remains submerged.

Economics is a tough taskmaster. So crime flourishes within a social ethic emphasizing possession, especially among those where possession is hard to come by. Our discipline of life is based on expanding production and consumption without end, and that system essential to livelihood can be turned off for anyone especially for those without a depth of assets, adding acute stress to alienation. We sought an outpouring from industry and so production propelled the vision that motivated people fundamentally throughout the twentieth century. But the urban world was seen largely for economic opportunity and a means for class advancement, certainly without a critical sense of what cities are or might become. In human terms, cities merely warehouse activities, good or bad. Cities could not be seen as active agents that might contribute to the human spirit, only as an environment for aggressive pursuits. And in the absence of a vision, cities became merely settings to be exploited as found, or used by homeowners to build capital gains.

And lost in the wake of building cities were the values of a once viable community, even as sentimental values of community linger in today's social psyche without a worthy understanding of what was really lost. The demise of the public philosophy lamented by Walter Lippmann also meant that urban philosophy could not grow or prosper, which left us with the notion that a public harmony was nothing more than a healthy market economy. Hence community values could disappear when confronted with the pervasive dollar values without too great a sense of immediate loss. So both the human harmony that community might have advanced and the ecological harmony that a broader public interest might have achieved had no independent basis to exist. Neither one had money motives to spur them on, so instead both generated first class crises at the onset of the new century. Both a creative lifestyle and a valid ecology have become, to use Toynbee's terms, the critical challenge to the creation of a higher, more civilized order, perhaps to the survival of civilization at all.

Imperative To Travel

Of all the activities associated with human advancement, the explosive consumption of open space, and the endless promotion of auto transportation are certainly among the most counterproductive and self-defeating. Consumption of land and promotion of transport are both essential to the American economy, its mythology and historic experience.

Evidence is close at hand. When the image of old crowded tenements like those built in New York after 1835 is compared to the idyllic manor houses of the English countryside—the historic model adopted by the American upper class, copied as closely as possible by the middle class, and concluded with the more humble tract houses that could at least accommodate all heating, cooling, and appliance systems plus a car or two. These gradations of class defined levels of the single-family detached house and spacious lot was also an urban import from our rural past, a kind of structure, space, and mindset that is, nevertheless, illogical in the geometry and function of cities. But the growing wealth permitted the sentimental rural image to prevail in the city, and it relieved people from being identified even distantly with the old, crowded, and decadent tenements of the crime-ridden inner city. Symbolically, then, the rural house beat out the profit-conceived, multi-family structure in dominating the urban form in a contest decided by money, prestige, and the lingering rural sentiment. And in areas where high-rise apartments put their excellence almost totally inside the individual dwellings, they made it possible to easily ignore the external environment. Nevertheless, the ideology of capitalism based on individual wealth could be better expressed in the single-family house, not in the compressed high-density "socialist" structure; if nothing more, it at least made possible a distant identity with rugged individualism.

Both the house and car were promoted vigorously by the commercial forces because *there is much more money to be made by urban developers and auto manufactures in independently built houses and cars than in building imaginative multi-dwellings served by efficient mass transit.* The entire picture speaks to the American mythology of individuality, private values, private consumption, private enterprise, and the vast panorama of the long-gone frontier. Thus the city was built without a sense of urban values or serious public interests. And where public values did appear, most forcefully in police and fire protection, these protective services grew apace with burnable wood houses and the protection of private property. Such limitations of the ethic and ideal are everywhere to be seen: in the roadways continuously testing one's endurance of travel; in the financial abandonment of older parts of the city; by the rapacious consumption of land and resources; by the urban environments that generate violence and crime; from the barren isolationist living of perhaps eighty percent of the population; in the sheer ugliness of strip commerce, declining commercial districts, and skid rows; and in the dominating physical brutality of streets, boulevards, freeways, masses of moving traffic, and vast parking expanses. All of these we know by personal experience, but it is the standard of living that is the prominent doctrine and dominant ideal, not cities; not the urbane delights of the mind; not public or private efficiency; and hardly ecological sustainability.

The growing scale of the house and lot and the growing number of cars each family musters to sustain mobility, like economics as a whole, can go on indefinitely—barring ecological burnout. The lot is good if one acre, better with five. It does not seem to matter that such acreage multiplied by millions of families creates environments of mobilized desperation, continually stretching scarce public finances to build and maintain the required follow up infrastructure, putting public band aids on social ills, leaving very little for public amenities or imaginative public environments. An important new museum or park is a rare victory. It did not matter that the automobile occupies over half of the critical downtown space and up to eighty percent of the land in shopping malls. The auto imperative both demands space and provides the means to both access and exploit urban land, leaving all who cannot drive—children, handicapped, elderly, the poor—in straits of utter dependence. But the infinite growth of consumption in sprawled cities nevertheless helps assure infinite growth of the economy. So, *from 1920 to 1970, the city became the possession of the automobile and the economy.*

Based upon the continued spreading of the city across the countryside, the auto promotion of scattered urbanism evades solution for even the most hardened auto-addicted person. Roadway expansion always leads to congestion, even stimulates it. The first hint of a successful traffic solution merely promotes greater auto development. *The meaning is simple: There can be no automobile solution for cities,* especially where sprawl and scatter demands utter auto dependency. Since there is no urban ideal to lead the way, the city becomes a development by destruction—which nevertheless assures economic growth. That destruction assures that the GDP grows apace with acres consumed, houses built, cars purchased, highways constructed, miles driven, fuel burned, accidents and injuries, not to mention the billions of hours wasted. So the city, being bound to the automobile, is effectively destroyed by the inhuman scale and operative demands of cars.

Created, then, is a technical and economic dynamism destroying human vitality, dominating behavior, consuming the environment, and controlling thought. A massive functionalism overburdens the family in work, commuting, services, and, not the least, the unprecedented material wealth. The issue is very clear: We can either continue to pursue consumptive technology and regressive economics; or we can seek a human dynamism in cities with an environmental grace.

Issue Is Human Purpose

The irony of destructive urban form is that specialized, pragmatic, direct response solutions—the stimulus to choose and consume only through the open-market process, with forced government construction of the necessary, though inappropriate, backup facilities—promote the giant runaway economy. Freeways are built, police are hired. So personal and public pragmatism rules and serves the forces of suburban sprawl that supports wasteful economics and generates forms of destructive runaway economic growth. That process reflects the virtual absence of principles to define urban efficiency, achieve ecological sustainability, promote social ideals, and establish a framework specifically for human development.

Moreover, while the city is expanded by the classic market process for short-term profits, the resulting urban form is the most permanent outcome of the modern economy. The city is built as if it were a consumable product while in fact it organizes human life for generations. And tragically we do so as the triumph of our growing standard of living.

Since we confine society to produce and consume, we blind ourselves to the greater human possibilities. So we fail to develop human organizations that might stimulate and sustain the greater social potential. That high scope of life could be the eminent role for community, seeking the varied, composite values of social life while becoming a major instrument to develop a universal efficiency, that is, a more rational economic structure of society. But today human purposes are subordinated to economic promotions, and these promotions result in the gross inefficiencies of people and governments.

The result is a cataclysm of modern living: the greatest source of the runaway economy, the wasteful time spent in congestion, and a massive new functionalism arising from the waste-demanding house and yard, cars, and huge roadways. Consequently, activities of high interest must now be minimized, like those in the arts and live theaters, or in facilities like parks and gyms. The winner is sedentary, passive, and exploitive television, especially for those who cannot ride or drive.

Plainly, a crisis of lifestyle accompanies the growing urban sprawl. There is alienation among us all and neuroses for many. Crime flourishes in an ethic emphasizing possession. Our discipline of life is based on expanding production and consumption without end, and that system calls forth the onerous crisis of human ecology and eternal shortages of resources we must face in our future.

Historically, we sought the outpouring from industry and so production was propelled with vision and undaunted determination. But a positive urban revolution was not sought; cities grew as a secondary result of industrial expansion, and so in the latter stages of output the increasing inefficiencies of cities resulted in an illegitimate urban process propelling economic growth. That growth resulted in mammoth gluttony, although it poses as new necessities. And as yet we have not determined what cities should be or do, except to expand private consumption and force public expenditures. Public values and civic achievements were suppressed in the ideology of private wealth and product consumption.

We wanted industrial output, so we mobilized labor, exploited resources, built cities, promoted science, and conscripted the universities. We even redesigned the universities and established schools for industrial purposes, as the "agricultural and mechanical colleges" created after 1862 attested. Corporations received huge land grants and mineral rights, were permitted favorable corporate externalities of industrial waste and pollution, and were granted the legal status of "person." However, it was in the form of the city, its free-market condition of growth, based on the real estate practice of randomizing human locality by the market's alienating forces that so confined urban life as we know it. Here, then, is the urban challenge, both human and ecological, that society must reconcile or face the dire if unknown consequences.

Great Spaces Demand Great Distances

Of all of the activities associated with human advancement, the expansive consumption of urban space and the endless promotion of auto transportation are the most counterproductive and self-defeating. Short of war, they are the most destructive combination of promoted systems yet devised. They represent a large part of both the seen and unseen propellants of the runaway economy.

Evidence is close at hand. When the image of old crowded tenements like those built in New York after 1835 is compared to the idyllic perception of manor houses of the English countryside, it is easy to understand why the single-family detached house and spacious lot expanded with the growth of personal income, despite the fact that some of the choicest living areas were of higher building densities close around Central Park. Moreover, the detached house conformed to the ideologies of the capitalist marketplace, stressing commodity exchange with private wealth, privacy, individuality, and class exclusivity, all that might be wrapped up in one escrow process.

The growing size of the house and lot could be infinite, as mansions testify, good if one acre, better with five, and with wealth a large acreage. It did not seem to matter that acreage by thousands of families eventually disintegrated urban form, putting a daunting burden of serving scattered developments with public transportation and services. But it turned urban transportation over to the automobile, which was made feasible by growing industrial productivity, expanding wealth, constructing costly roadways, and accepting the burdens to pay any price for the escape to the suburbs.

The spread city often multiplied commuting burdens and commanded huge subsidies for highways from government and great payments for parking spaces. The century-long struggle to construct urban roadways and the demand to build parking spaces at both ends of every trip has yet to be resolved even to a minimal satisfaction. It did not matter that the automobile-dominated half of the critical downtown land and eighty percent of that in shopping malls. Always to expand facilities for the automobile was accepted as an imperative for progress; never to be questioned in principal or for its human outcomes. The only issues were limited to setting priorities and finding means. *Never is the question examined far into history or far into the future,* either of which might have given a clear perspective as to what was really happening to the whole city as a livable, vibrant place of civilized pride. The true imperative was not for progress but rather for the creation of a desperate dependence upon machines that between 1920 and 1970 made themselves absolute necessities to obtain everything in the city while they destroyed the best of what the city could have offered.

Yet auto possession of the city today evades solution, for the solutions we sought have become the problems. However many roadways are expanded, congestion follows, every successive round of expansion reinvents futility; for each round calls forth still another round of destruction by development. The auto solution—the foundation for sprawl—feeds an endlessly growing necessity, and demonstrates once again that there is no urban ideal by which the force-fed auto-mobility might have revealed. Exposure of this huge historic fraud could have been found in (a) the human imprisonment in both the isolated residences and cars; (b) a combined examination of personal, governmental, and even corporate expenses; (c) comparing vehicular and human environments; (d) examining the urban problems arising from sprawl and automobility, including health, education, and crime; and (e) evaluating the sources of the runaway economy. All in all, the more wasteful the runaway economy becomes, the more its figures reflect misshaped growth, especially the extra cars, accidents, and conversion of the city to absolute mobility. So the city becomes a nearly perfect means to expand the most fateful side of economics, its runaway force.

The issue is clear. Whereas the city is inherently a close relationship of human activities—living, producing, servicing and associating—and social interests like recreation and entertainment, the American habit is to stretch the spaces to their maximum feasible ability, destroying the natural efficiencies of proximity the city once had and the grand, varied common open spaces it could have had.

That efficiency is defeated by an interplay of oversized private spaces only partly utilized and the least efficient form of transportation, the automobile. The inefficiency of one promotes inefficiency of the other while reducing the city's ability to serve a far wider range of human objectives.

Created, then, is a massive functionalism never before seen. It endlessly compounds working, shopping, commuting, servicing, and repairing the supposed largesse of encumbered spaces and distances. Cities are no longer the possession of people. Rather, they are the property of corporations that build the massive urban formlessness and the materials and machines to keep that formlessness minimally functional, all in the name of the high standard of living.

Promoted Expansion

The lack of an historic and philosophic focus upon the city as a basic and unified theme is prophetic to the still unrecognized defeat of the city as a valid social and reasonable economic instrument of society. What we now see is an inordinate new dependence upon machines, a dependence masking the massive spatial and material inefficiencies, and the bleeding of humanity rationalized as wealth. *The city has become a public arena for expression of private values nakedly expressed as the promotion of production and consumption without a counterbalance of positive public values.* This corruption must be suffered, it seems, before reform can be undertaken. There is good reason, therefore, to say that the city is historically a non-entity, merely a collective harboring of the "real" facts of life.

That manifestation began, simply enough, in its gridiron urban origins. No other arrangement seemed so natural. It could be universally applied, was endlessly expandable, and it accommodated virtually any activity admissible to the city. It was manageable without much management at all; and set forth the image of democratic equality. No better plan could have been devised to meet the unknown requirements of industrialization and rapid urban growth as flexibly. Straight streets and rectangular blocks could be set and surveyed without agonizing decisions. At the start, the intersection of two roads in the wilderness would do.

The very idea of the urban grid pattern promoted the endless growth of consumerism. And since much new wealth was devoted to plumbing, electricity, heating, air conditioning, and appliances, the technical obsolescence of the old dwellings accelerated, especially for a place without a driveway and garage. The car's impact was decisive. Auto production barely existed in 1900, producing a mere 4,192 vehicles. By 1905 output reached 24,250, then 895,930 in 1915, and 3,735,171 in 1925. Thus, a massive industry exploded in twenty-five years, and production suggested the scope of the urban impact after World War I. Wide boulevards strained urban development budgets after 1920, followed by freeways mostly after 1945 when their enormous costs became the dominant struggle. Off-street parking lots and garages accelerated, becoming legally embedded in zoning ordinances. While in the 1920s people complained bitterly about the city's inability to accommodate the automobile, by 1970 the auto conquest was essentially complete, people nevertheless continued to complain about urban traffic and parking. Since then, traffic and construction continued the automobile's expansion while some people are beginning to see that there can is no possible relief from traffic congestion and that the sprawled suburbs are the crucial forces of that debacle.

Hardly surprising, therefore, is the fact that every major urban innovation reflected the buildup of auto dominance: the roadhouse or tavern, drive-in theaters and drive-through eateries and banks, supermarkets (now super-stores), discount department stores (Wal-Mart and its competitors), shopping centers and malls, and industrial districts, and especially the long boulevards of strip commerce. All of this, of course, was in addition to the auto's own service stations, repair garages, dealers, and wrecking yards. The greatest innovation, of course, was the freeway, billed as the only roadway equal to the performance of automobiles; and it is that until traffic reaches its capacity, at which point it grossly breaks down. So today the traffic engineers are desperate to maintain freeway capacity; thus they have introduced the diamond, multi-passenger lanes and the signals that limit the traffic entering the travel lanes.

It was Norman Bel Geddes who designed the General Motors exhibit at the 1939 New York's World Fair and who envisioned the city completely transformed for motorization via freeways and effectively propagandized smooth-working mass movement. That city was a system of nonstop motion, which the public essentially bought, hardly recognizing the implicit inhumanity and drastic ecology the plan mandated. Yet it was only thirty years before this technocratic fact was indeed established. Geddes published the exhibit in a book, *Magic Motorways* (NY, Random House, 1940).

Freeways were revolutionary to the city far more than revealed by merely their traffic. They cut through the flesh of neighborhoods without regard for what humanity they contained and helped destroy established neighborhoods.

Then, in moving to suburbia, and having those freeways, most people believed they had reached the ultimate condition of progress, the idyll of both town and country, yet were destroying both. But many now belatedly realize that something is very wrong in buying into the bedroom suburbia, by its design a sterile urbanism. So also, it is astounding that Americans continue to build exactly the same kind of dwelling in both the country and the city, a barren structure voiding the very human excitements in daily civilized affairs, but that dwelling carries no greater vision of behavior than advancing consumption, that is, acquiring the toys of the advancing standard of living. When collected by the hundreds of thousands, this land amounts to a huge urban consumptive destruction of not only urban land with its forced mandate to maintain an acceptable landscape, but also the human scale of creating a magnificent diversity of urbanity is totally lost.

We need to recall again that an effective city is inherently a quantity, quality, variety, and intensity of services, activities, and interests directly available to everyone. Most of all, those conditions are possible only at the human scale where urbanity and humanity are united in common and integrated form. *Whatever exceeds a human scale diminishes both urbanity and humanity.* This tough lesson needs to be learned through a hundred design steps and for a hundred benefits arising from the human scale.

Tragically, while the history of cities is to build, gradually improve, and then build for urban magnificence, the Americans have built their cities and then abandon them by moving to the suburbs. Improved urban potential is sacrificed in the escape to sprawl. The wealth that might have gone into improved urban qualities is given to vast spaces, distances, and transportation. An attractive public life is forfeited to isolation in the house and car. Little wonder that an American consumes three times as much energy as a Frenchman.

Destroying Freedom

The American Dream is silent on cities. Americans didn't really want to become a nation of cities. Yet they wanted industry, and cities were required to bring all of the manufacturing ingredients together, especially labor. Americans made sure that there was economic opportunity. But there was little to make them aware of the urban foundations of freedom. In seeking production they built cities enthusiastically and this enthusiasm set in motion what would become economic determination. Before 1900 we promoted industry, despite their ugly smoke-filled conditions.

In the following half century, housing improved, water became cleaner, and the desperation of old living conditions measurably eased. Most importantly, the city became electrified and natural gas was introduced. These led to very substantial improvements through a mechanization of the house: stoves, water heaters, refrigerators, irons, washers, radios, mixers, improved lighting, and central air conditioners. Life, especially for the housewife, became notably more attractive each decade.

The postwar baby boom intensified urban growth, which was spurred also by labor freed through increased mechanization of farms. Houses were designed upon larger lots to more completely accommodate the technical infusions, and added the two-car garage and a lengthy driveway. The huge subdivision tracts of identical houses of close-in suburbs of the 1950s gave way to larger and more often custom houses on ever more spacious lots in the more distant suburbs and scattered rural districts. Thus opened the era of long automobile commuting on congested gold-plated freeways and forced cities to fight to maintain minimal mobility. In but two generations mechanization dominated the house and city.

The central fact of the city today is that it presents an image of freedom, but works serves mainly a penetrating system of gluttony. But what can work? The awesome challenge today is that we have neither the theoretical nor practical foundations to build truly efficient cities, compose an urban environment that is ecologically sound, shape the urban environment with social grace, and solidly implant human freedoms that should be fundamental urban rights. We have yet to recognize that cities themselves constitute the forms and functions of better living, inherently in themselves, as well as what they contain, like our body as being more than the organs that constitute it. Instead of perceiving cities whole for their total impact, we see only the parts and pieces, the houses and cars, and the problems of giving them full reign in the city. We must become aware, however, that *movement of personal machines cannot be a central function of cities, and trying to incorporate them into cities is the greatest urban misfortune of the twentieth century.*

Consider that degenerative process as it governs present urban development. When our system of wealth encourages the most spacious properties one can afford and the most space-demanding individual transportation one can manage, the urban form is exploded. The greater distances demand a fundamental takeover by cars, especially because public transport cannot effectively serve low usage over great urban expanses, even less so when travel destinations become widely scattered to work areas, shopping, and entertainment. But those distances require that *urban* children must be bused to school. Postal, police, fire, and utilities must command fleets of vehicles to span their stretched services. Yet this urban form determines how and what we produce, how we spend our money and manage our time, and effectively tells us what we can and cannot do.

The city thus determines the most basic features of modern living. But rather than elevating the content and quality of living, the city becomes our greatest burden of economic and social cost, determining how we organize our private lives, diminishes the effectiveness of schools, reduces the condition of health and longevity, and forms aggravating nests of crime. The huge garbage dumps and wrecking yards tell their own story. Yet this dismaying tally of life remains on the accounting books as a high standard of living.

The conservation movement pleads for voluntary restraints to painfully address ecological sustainability but has yet to discover that the problem is deeply structural within the urban form. Most of all, *the movement has yet to realize that the gravest problems of cities and their greatest possibilities are essentially the same issues, both crying out for common solutions emphasizing broad and positive strategies.* When that case is demonstrated, we can look to a new order in the way progress is imagined.

The current debilitations call both our humanism and our freedom into question. Long ago Robert Heilbroner signaled a future of shortages and constricted behavior if we continue the sleight of hand of viewing *economic* progress and *social* deterioration as unrelated phenomena. Yet, even without having to directly face disruptive shortages, the course of urban decay now presents an appallingly clear and present danger of a new, greater tragedy of human development.

Contradictions

The irony of destructive urban form is that the pragmatic solutions creating the giant runaway economy are precisely the problem. Not only is waste the problem, the complexities of waste create utterly inhuman conditions of life. That reflects the virtual absence of principles to define urban efficiency, organize ecological sustainability, promote social ideals, and establish an ethical basis for a sound economic and social functioning of society.

Since we organize society overwhelmingly to merely produce and consume, we have lost sight of inspiring human purposes in society. So we discourage the development of stable human organization that might conceive and sustain greater human purposes. But today broad human purposes are subordinated to the market and economic promotions and, tragically in the pursuit of progress, urban form offers the most effective means to promote the runaway economy.

The irony embedded in the urban sources of the runaway economy is found in the terminology of urban development. *Land use* emphasizes what is used, not the purposes of urban form. *Central business district* stresses economic domination, not the public values of the heart of the city. *Sub-urb* alludes to the urban sprawl of development. *Zoning* refers to the segregation and isolation of urban form, not its integration and unity. And, too, we must stress that zoning is based on a negative vision of the city, having been established solely as a means to separate filthy, smokestack industry from living areas. *Subdivision* similarly stresses the separation of private interests, not wholeness such as community. *Transportation* stresses promoted mobility, not an efficient balanced access achieved through unified design and public transit. *Housing* emphasizes only the bare necessities of living, a notion not far from the barracking of people, certainly not an organic form that dwellings might constitute. Meanwhile we invert the meaning of key urban terms. *Center* is often used inversely to the degree that centrality is lost. *Urban* replaces the idea of a city with its historic ideal of unity of social, economic, political, and cultural life—behaviors signaling a vital civilization. Similarly, the word, *community*, is itself now used commonly as a sentimental reference to almost any identifiable group, now after true historic community has disappeared.

But it is *population density* where the greatest misconception appears. Yes, where tenements existed without adequate interior spaces, without amenities, and without exterior space, high density may be compared to experimental rat crowding, where there is no physical space to move about. A high-density community building can, however, reveal an unprecedented range and variety of interior and exterior open spaces with no congestion whatsoever. Yet the comparison seems to provoke fears of a modern black plague; there is no basis for such a fear. Indeed, in the modern setting of affluence, as we can all see, congestion is completely vehicular, arising with mass travel, not living spaces. As affluence grows and cities become more wasteful, the human benefits become ever less meaningful and lost.

The basic revelation of these several terms is that cities are not seen as wholes with human efficiencies and amenities, so ideal forms are not in our conscious awareness. Nor are the broader range of human interests brought into focus. As affluence has grown and cities become more wasteful, the human benefits of urban form diminish sharply. Rather, the effects of great wealth on cities today merely makes roadways and their accompanying facilities more urgent, drains public finances, and accelerates a race that can never be won.

The central issue of cities is responding to human purpose, which we must be alert to continuously create. That purpose is best assured by a system of social goalmaking, as I will show. It also becomes the chief means to break away from the erroneous convoluted illogic of reactionary pragmatism we use in always building more roads, more extended utilities, more fire and police protection, and expanded war on crime. Each is a specialized response to important issues but causes a superficial and false format when merely promoting economic growth. Community, however, is capable of creating that broad human purpose and becoming the medium for goalmaking to continuously raise human value in society, supplanting the monolithic pursuit of increasingly counterproductive economic growth.

First Principles

We see in vehicular congestion arising from very low population density the *negative* paradox of urban development, a congestive paradox of too much. Too much of anything is a human burden, like too little. But too much space means too much travel, too much machinery, and this results in too much cost, too much management, too much repair and service, too much wasteful behavior.

A series of urban development principles can reveal true human benefits to achieve a *positive* paradox of urban form. That principle is simple but dramatic. *Urban spaciousness is possible only through compact development.* The lesson is to gather dwelling units together, with ample or generous interior spaces, so that the oversized and underused personal mini-spaces of lots can be in effect gathered together for highly varied, large-scale common spaces close to one's dwelling. By so doing, immense spaciousness is reserved not only for parks and recreation, but also affords huge spaces for "urban farms" and "urban forests," botanic gardens and ecological study areas. These most generous spaces, open to all, nevertheless require but half of the acreage of today's average urban requirements.

From this alternative use of urban space another major principle is derived: *Only with compact urban development at the human scale can there be a universal urban efficiency.* Such efficiency can radically reduce the gargantuan and mostly useless functionalism generated in the twentieth century. Personal burdens are reduced. Commuting sometimes can be virtually eliminated. Families can dispose of surplus cars, possibly all. Costs of utilities and government can be significantly reduced. Even business itself can reduce its overhead. Cultural facilities, health services, schools, shopping, and much local employment can be assured only minutes from everyone's door.

Compactness of development in urban nodules permits a third principle to arise: *Transportation can be rationalized first to minimize its need, and second, to increase its efficiency.* This principle shifts the emphasis from building transportation as an endless necessity by the least efficient form of urban travel, the automobile. Rather, *the principle greatly increases mutual accessibility among urban activities through integrative design.* Activities that are related to each other, whether public or commercial, are placed together. Indeed, all activities in compact development are only a few minutes from one's door, a few seconds walk between them, and never more than among stores in today's malls. Transportation can then serve after the limits to access are reached through design. Walking is restored in the close environment where it is the most efficient system possible (possibly with about as much walking as required today in getting to and from a parked vehicle) while also renewing casual association and personal relationships within urbane environments. Transit will then be fast, efficient, and frequent, operating between urban nodules. Automobiles will fill some special travel needs but mainly serve recreation.

The system of compact urban development then achieves a fourth principle, following spaciousness, efficiency, and rational access, possibly the greatest of all benefits in the long term: *The compact nodules of urban form coincide with the integrated form of community.* Thus physical and social principles coincide with a human beauty. The intimate and effective community is the people's own institution, being local, comprehensive, and integrated for only their interests. The unity and clarity of the community defines and serves the unity and clarity of its organization of human affairs and the perception of positive possibilities. Clarity is both a revelation of the deeper human problems and a basis for inspiration of our greatest potential, today the fatal inability in the city's formless expanses. *With clarity of the community's size, boundary, center, facilities, roles, capabilities, and the people's interests and goals, it can serve all of those interests to their optimum.*

Over the last century we have seen how the rationalization of production progressed rapidly with the principle of the production line. More recently we have seen how technological advance has employed computers to completely automate many of the processes of production. With community, comparable advances in directly serving people can occur, which can then organize the means to achieve the larger, positive range of human freedom.

A few important matters must be kept in mind. First, if people are to truly matter, we must keep them uppermost in mind, as well as embedded in the constitution of community. Second, we need to reconsider what is public and what is private, for today's overwhelming stress on privacy has added seriously to the debacle we live by today, for that monolithic fact of urban life lies close to the root of the present denials of both personal freedom and privacy. Third, an integrated urban form cannot arise as the result of free market behavior, for the market inevitably responds to only money values and prompts a socially irrational dispersal of behavior, a contradiction in community. That system, for all of its value in supplying consumer goods, cannot integrate for social and cultural, as well as economic, functions, that is, to the larger range of human purposes. That is what has led to the distortion of deeper human values and lost social vitality, quite in addition to wasting our resources, our time, our temper, and the most central possibilities of specifically *human* progress.

A rational urban form leading to a least-means economy and then to an expansion of social freedom is, I believe, close to what we must strive for. The significance is not endless growth but rather the expansion of harmonies and human possibilities throughout society. This means more than anything a focus on human qualities, not output, not wealth, although these are supportive. The goal is a system that inherently empowers the person in daily affairs throughout life, with each person determining his or her own terms of social engagement.

Over the last century science and industry have established a massive new foundation for life in society and a great exaltation of life. Most importantly, that now must mean a freedom from the classic confinements of economics, from the struggle to get on or the problem of money management. Technology and industry will continue to develop. But they cannot continue to dominate life. That is the huge lesson we learn from economic determinism and its massive result in the runaway economy. We will be both freed and challenged anew to pursue an expansive life with the same intellectual vigor and determination as propelled the Industrial Revolution.

Social Challenge

In Toynbee's terms of the stimulus and response by a civilization to its challenges, our challenge today for the civilization we are creating is that the corporate industrial system we now live by overwhelms both nature and society. This means our challenge is within the social system, within its own set of powers and manner of thinking. The critical challenge lies within our social self-perception and creative understanding of our self-development and social dynamics. Today's crises of society ultimately rest within how we think, organize, and act. So the conditions we face—and the critical challenge of our era—exist at an order of magnitude above our normal manner of thought, creativity, initiative, and belief.

That challenge of breaking today's orthodoxy is similar to that which resulted in the Reformation in the sixteenth century. Therefore, what is required is a massive and fundamental paradigm shift, like the *environmental* movement has set out to accomplish. But that movement cannot come to fruition until it is joined by an even more basic *social* movement, which will allow it to rewrite the terms of behavior now reflected in the economic domination of our social existence, by which we can whittle down the runaway economy for both realms of social and environmental terms of our civilization.

Our problem is further complicated by the fact that today's society and its driving forces are snowballing the terms of further growth of economic power that builds upon the enterprise's endlessly increasing profits in an infinity of insane acceleration. Today's system is protected and disciplined by a clutch of ideologies and political forces rising upon the many capitalist triumphs. The power, speed, and repetition of current stimuli and responses of enterprise cannot recess to take stock, nor is it inclined to do so in the grip of its blind pursuit. It is at that point that both social and environmental objectives must fully confront the politics of corporate economic power.

By a growing concern, tragedy is written into the present course of society, a tragedy rooted in a too-successful system with its own built-in and circular stimuli and response, a system that mattered little in 1900 but mattered overwhelmingly by 2000. Being the first ultra-successful society, a radically new stimulus must now present itself; yet we have no experience to serve as lessons. But everyone is in it, everyone a potential victim, and everyone a responsible actor.

The very real struggle today is like the challenge of civilizations described by Toynbee. He notes how an "ailing civilization pays the penalty for its failing vitality…[and so it] rules with increasing oppressiveness but no longer leads." That is our case today. But for the first time in history the affliction is not a failing vitality, but an ultra-vitality endangered by both ecological and social burnout. A rising "proletariat," argued Toynbee, "responds to this challenge by becoming conscious that it has a soul of its own and by making up its mind to save its soul alive." (*A Study of History*, p.77) Our all-but monolithic tradition of narrow economic stimulus and response today is, not just in business, but also in government where paranoia resides in keeping the economy growing, and with the people who are trained in the mindlessness of our old manner of work. That is where our great paradigm must shift.

Cities are crucial of course, all the more because in the past they have not been favored with human visions and inspired social action. They could be said to be the heart of it all, being both the core of human ecology and the organizer of human behavior. Therefore *a human renaissance shall be an urban renaissance.* And the heart of humanity in cities is how persons can form themselves for greater experiences of life, that is, within the diversity and richness in which human life is first cultivated, and then given the fullest opportunities.

CHAPTER VI

ECONOMIC DISCONENTS

Power

We have seen how the urban discontents of life derive from a lack of social understanding of cities, inadequate urban development theory, misplaced organization, and misused powers to shape urban life. Consequently, cities have become the sinkhole of democracy.

The discontents of economics, however, derive from the opposite conditions: wide and intensive development of operative knowledge; highly developed theoretic system; doctrine of education and ideological indoctrination; powerful organization at all levels of society; and both a unity and diversity of freely-used powers. Moreover, economics deeply penetrates society, including local and national politics, career development, promoted consumption, science, technology, and bureaucracy.

The economic presence is not merely a service to society; it constitutes the greatest collection of society's special-interest groups and reflects an overwhelming concentration of social wealth. The reach of those special interests is phenomenal, whether one considers its global, national, regional, or local powers; and whether political, social, and cultural arenas of society. That reach also dominates society's resources, including capital, labor, minerals, land, the core purpose of education; or the content of research, promotional penetration of personal thought, character of the social mind, market control of urban development, and extensive control over the legislative process at all levels of government.

The varied powers of the special interests of economics derive in part from the many forces of capitalist ideology. Well honed from a seven-decade contest with communism, today's capitalism basks in triumph upon its world victory, displaying vast versatility, incontestable productivity, commanding economic determination of social behavior, and its political control of society.

That ideology stands upon at least four major foundations: the concept of corporate free enterprise as an organizational principle; free markets as a self-correcting process; a large, diverse, skilled, pliant, and a reliable labor pool; and money as the universal investment tool, medium of exchange, and end goal of business.

The enterprise's impact upon people is that it not only promotes urban growth to concentrate workers for the needs of the ever-growing economy; it also requires that workers be fragmented to be effectively organized and process the skilled labor needed for factory, laboratory, and office. To maximize dependence and loyalty, businesses frown upon organizations and associations that might compete with the work discipline they desire, such as those that might distract workers from relying completely upon their wage work.

The village community once afforded a cohesive social world in which work activities were embedded in the larger affairs of the immediate family and village. But in the structure of booming amorphously organized cities, work was largely removed from its social context, thus elevating economics to a supreme position; and social life was reduced to a dependence upon employment and its inherent risks. No longer were life's activities unified in one house and community. No longer did social organization prevail in life. Economic interests then gave their central direction to other interests, thus assuring that workers would always follow the dictates of the job.

That rule of work established the first of two powerful forces behind economic determination, first, the complete dependence upon money income for the *production* of all of life's necessities. The second force completing the economic determination appeared with the power to promote *consumption* among the psychologically fragmented population. With essential control over people's role in both production and consumption complete, the transformation from a communal direction of economics to the economic determination of social behavior was complete.

Therefore, it can be argued that capitalism essentially substituted economic organization of life for that of community. Most simply, society was reorganized singularly around production and consumption. That new imperative modified a person's life career to accommodate the growth of cities; the organization of society by large, powerful corporations; the way people think about their life; and to their newly internalized, money-dominated social values. As it displaced community it established over several centuries the conditions required for a thoroughgoing economic view of human nature, views that penetrated the unconscious base assumptions of life. The loss of community decidedly confined people to being producers and consumers, thus contributing to their lost personal empowerment. It meant that the *soft* qualities of traditional social life became subordinated to the *hard* powers of modern economics.

In the end the tradeoff largely substituted personal wealth for the social worth, as part of the loss of extended family or community. While doing so, it created a vast cosmopolitan life never before experienced by society, a rich and diverse cultural participation for those who appreciate it and pay for it. People could enjoy new forms of entertainment, recreation, and sports, especially the whole new class of robust outdoor water, winter, air, and land sports, including mixed feats like the triathlon. These activities usually centered upon what people could purchase.

If there is ever to be a modern rebirth of community, it can reestablish an essential social unity and vitality that existed in the historic community. Community might then again play some significant economic roles, including involvement in public and private services such as retail and professional; and popular involvement in the arts, recreation, and other fields of human enrichment. Most importantly, community can achieve unprecedented efficiencies of its community building, second, by helping to develop a rational system of urban transportation, and third, in effect, becoming a consumer's union. Once again, many social affairs can then take precedence over today's economic domination.

In its traditional terms of seeking to end the human history of basic poverty, the current economy has succeeded mightily (while maintaining a population in poverty). However, the economy also continues to seek the endless growth of national output as an extension of the goal of continuously expanding corporate profits. That objective, I believe, results in today's historically unique economic determination of society. Our purpose, therefore, is to establish a new social balance while creating a new range of personal freedom.

Need for Redirection

The endless goal of expanding corporate profits driving society's pursuit of infinite economic growth promotes two emerging public tragedies of our time, the dehumanization of our social system and the destruction of our ecological foundations. The continued growth of the GDP promotes the already vast runaway economy, a destructive material redundancy and the basic source of both social and ecological debacles.

Economic growth all too clearly reveals the great corporate game of social and political power, based increasingly on runaway economics that also concentrates wealth for the economic elite. That is, economic growth structures the economic plumbing to enlarge the money flows into the capitalist coffers. That game of power channels money flows like the fantasy board game of Monopoly. Without limits, the fantasy accelerates the run on land and resources, subjugates people to the dictates of economics, and keeps tight reigns on government and its "political purchasing power" within the legislative processes.

The two debacles of society point to the failure of the highest levels of social leadership. It does not clarify philosophically, for example, the distinct roles of each stage of social development. Nor does it enter into public debate the critical issues of fundamental change. It reflects all too singularly the narrow scope of business leadership that avoids thinking at all about the greater issues. We need to learn that the greater the impacts of change, like those we are now experiencing, the greater are the needs to understand and direct those impacts. *Today, I believe, there is no greater need than to understand the dynamic tectonic shift from a system of general human scarcity to a new system based upon widespread abundance.* That shift is not only fraught with the grave issues of basic social and ecological validity. It is also laden with human promise, an outlook opening a new social destiny.

The philosophy I refer to is, first, to understand and expand human purposes to reflect the new widespread wealth and, second, to understand the enormous implications that such social change might bring to fulfill the new, broader, more promising human purposes. The unfortunate fact is that even with the last two centuries of profound social change affecting the essence of what we are as human beings, *we don't have a useful tradition of giving conscious direction to the forces that so deeply impact themselves upon our lives.* The only evident exception is our experience in perfecting technological and economic performance, but that "expertise" of economic development is closer to the problem than a valid social solution.

So, what are some of the foundations for a meaningful social philosophy? Taking a great leap concerning my assertions, we might consider the following points: We must first greatly moderate our impact upon nature that we are now beginning to understand. Second, we can continue to encourage higher human aspirations in the classic arts, the range of recreation and sports. But these are the easy answers. The more challenging and more important questions get closer to basic human purpose and social organization.

Undoubtedly we need to shift education from career skills to learning for lifelong personal enhancement, to help each person expand her own most meaningful experiences in life. This means that the more basic philosophy of society needs to encourage that experience which each person defines and pursues for himself. And a new personally defined range of experience will mean that *our greatest goals will revolve around the creation of a new vision of freedom itself, goals that are personally expansive, bear deeply within one's own personal imprint, build upon associations, take personal initiative, and seek deep open-ended interests.* The greatest human promise is inevitably person centered, regardless of future technical advances, and it is for personal development that we can look to for the greatest human creativity—the creativity of themselves within vital human association. Such a course implies developing the broader psychological and social basis for freedom, no longer dependent upon the paradoxically confining economic terms of life.

The growth of human purposes becomes an expanded form of freedom we can learn to build upon. *Human freedom itself becomes human purpose, no longer merely a protective freedom, a freedom emblazoned into the deepest human purposes at the core of social behavior.*

As it has been society's age-old objective to break the bonds of the *necessity of nature;* now we can begin to break away from the *necessity of economics* as well. We can build many avenues of human cooperation and reduce the severities of competition. Then freedom can grow as a heightened social ethic.

System Without Limits

This book notes several little-noticed conditions that have appeared in society in the last half of the twentieth century. Arising with general affluence appeared the *Commercial Revolution* that grew out of the Industrial Revolution, which in turn led to the growth of *economic determination* through the strategic powers of the corporations with their deep penetration of personal and social behavior. That resulted in the rise of the *runaway economy.* These three resulting conditions reflect not only the huge success of capitalism but also the larger effects upon society by the first general appearance of economic abundance that broke the old chokehold of biting scarcity. They reflect the triumph of capitalism but also herald major issues society must yet address if the outcome is to result in an uncompromising human advance.

Freed from historic scarcity, the new wealth nevertheless severely confines human behavior. Society's new abundance unfortunately stimulates the great wastes of urban sprawl, the ravaging urban rule of the automobiles and freeways, and the lavishing promotions of the giant new scale of advertising afforded by the wider price margins of luxury goods. The new purchasing power of consumer's freely expendable wealth translates into a corporate power that directs wealth into endless economic growth. The seeming freedom to choose from the vast range of new products gives the corporations great power to direct that "freedom" for their own advantages.

The Revolution of Commerce was really a revolution of wealth, for by converting industrial capacities into commercial power, a force has arisen that is not yet recognized widely by society. Simply put, consumer freedom in reality presented corporations with the freedom to dominate consumers. Therein exists the growing economic determination of social behavior and the increasing power of the self-serving corporations that generate the destructive runaway economy.

And therein lies the potential for social disaster, the built-in acceleration process that is imperative in the competitive free-enterprise system. Without external limits or social restrictions, there appears to be no way to moderate the acceleration. The system itself is incapable of taking into account either the numerous social crises or ecological devastations. If every enterprise must compete, it must do so without limit because the game is ruled by competition. And that mindset goes to the heart of the economic science. Economists seem to have observed competitive enterprise behavior and dubbed their observations as scientific truth, and only a few have noted the adverse social repercussions or progressive ecological damage. Adam Smith could never have foreseen an economy racing to infinity. But today that is the apparent race to ultimate disaster.

So, in this structure of action, the system promotes the excesses of gluttony as a central aim of economics and a great value of society. No excess is recognized. Thus, by its very aims, the system is out of control. This dominant institution of public organization is self-seeking, endless in its grasp, and has converted the profit motive into an immense game of economic and political power. Consequently it views all efforts for a rational dialogue as interference to free-enterprise, and couches the system as human freedom. Non-economic goals are suppressed or reduced to political impotence. No ecological damage is accepted, unless there are profits to be made through "corrective" programs, thus creating new "negative" resources for growth.

All power is dangerous, but when power is built into a system of economic growth it becomes social corruption. This mania for endless growth controls not only the material content of society but also the substance of education, the behavior of individuals, and the psyche of the person. That system can then effectively veto legislation that might underwrite corrective social or ecological measures or propel a promising social evolution. But growth is the driving doctrine of economics and for that end it has co-opted both government and the public mind. One only has to listen to politicians to perceive their anxiety to maintain steady growth. Hence this system of corruption rules, and rules almost absolutely.

The revolution of wealth exposes another side of the advanced economy that is necessary for the development of valid and vigorous society. The concentration of massive productive wealth raises an issue quite beyond the old question of distributive justice. That is, while corporations dominate the producers' and much of the consumers' sides of the economy and maintains their grip upon the public-interest economy, the public is both strapped for essential funding and the disparities of wealth work against the varied forms of social cooperation that are essential to avoid crises and promote a progressive public system.

Other Economy

Today's revolution of wealth exposes another side of the advanced economy that has had little attention but is absolutely crucial for the development of a valid and vigorous social development. Control of the new wealth effectively raises anew the old issue of economic justice, and does so quite beyond its historic question. That new wealth pits the power of the productive and commercial side of economics against all consumers, since further economic growth depends upon promoting maximum consumption. Thus the old question of distributive justice must now stand against the new profligate wastes of the runaway economy. It is a system of using wealth to promote wealth, even as much of the new wealth is a specious waste while being denied to those remaining in poverty.

We can argue that the consumers' economy is a sector that should perform a role that is an equal and opposite condition to the suppliers' side of the market, a role that goes far beyond choices and purchases. When producers can—by multiple and powerful promotions—deeply influence the buyer's behavior, they interfere with and to a degree control the free market that is fundamental, if not sacred, to the system of capitalism and economic analysis. The issue could hardly have arisen when scarcity generally prevailed; but when affluence arrived and freely expendable money appeared in force, then the issue of supply-side power became acute. So now that shift has created a new power of economics and raises very basic problems of human justice and freedom in a society largely controlled by corporations.

That power was in the hands of consumers when they were struggling to raise their standard of living. Then the purchase was simple: buying reduced the decision to make, model, and price, with service and quality being secondary. Basic decisions were then controlled by need; only the margins of product selection were available to most buyers. But when affluence took hold after the house was filled with its range of needed appliances, then the buyer's seemingly wider range of product choices gave the marketing arms of corporations vast new leverage in promoting more automobiles, bigger houses, more expensive clothing, luxuries, and travel. And then a very new avenue of growth promotion took hold, made permissible by the wider price margins this new level of goods afforded. The supply-side economy could then expand its power directly into the consumer-side economy through its product promotions, especially advertising and credit; thus established were new supplier controls on both sides of the "free" market. Very simply, the invisible hand lost its touch.

But, for what remained visibly evident, the competitive free market continued to prevail. But the dramatic growth of advertising and credit in this new realm brought into the market equation a corporate influence—or interference—into the consumers' own bailiwick. More powerfully, the forces of urban sprawl built into the consumption process new arbitrary demands for consumption: large-yard equipment, longer commuting, more dispersed services, and the necessity for more cars and more continuously demanding functions of upkeep. What was being built was a hardened *structure of consumption,* a created necessity that appeared like the former necessities of subsistence associated with the early growth of the standard of living, but in fact was really a different sort of economy reflecting new commercial power. A parallel expansion of the new necessities was the public requirements for new highway construction and the greater costs of more dispersed utilities, schools, police and fire protection. The new structure of consumption was a promoter's paradise, elevating corporate control over both sides of the market, and promoting further economic growth. This growth strengthened economic determination and increased its chief outcome, the run-away economy.

When abundance appeared, no feature in the market filled in for the discipline that once existed during society's early struggle for necessities. No new authority, informal or formal, filled in where the invisible hand noticeably weakened. But the old necessity imposed by scarcity and governed by demands of nature gave way to the new necessity of economics. And so supply-side economics smoothly filled in the "vacuum" once occupied by the old scarcity discipline. Thus, upon the new abundance, society over merely a few decades entered into a hyper self-generating system of corporate promoted economic growth beholden only to the triumphant free enterprise system.

But with businesses operating on both sides of the open market, economic growth occurred more completely from practices internal to the system, practices responsible only to the system itself, even as growth increased the threat to eco-logical non-sustainability. The suburban isolationist crises merely prompted a greater dependence upon buying the new necessities and pleasures of life. While a substantial population remained in poverty, the increasingly wasteful form of the city, plus the increasingly intense advertising to consume, resulted quite hap-pily for business as the new foundation for economic growth. Consequently the national government itself remained morally in the hands of the corporate system by pursuing economic growth without an end in sight, thus sustaining mono-lithic economic determination while expanding a level of an ecological draw down unsustainable for the world's population, if not for the American. Thus both the people and government have been arbitrarily pulled into economic expansion through a charade of wealth working against basic economic, social, and ecological rationality

Business cannot dismiss its monolithic mandate of pursuing increased profits; nor can the economists find a basis within their theory to respond to ecological necessity, social issues, or common public aspirations. Only its market system processes bind the system. Thus the span of human purpose is blocked out of both business and economic analysis.

What we live by, then, is not an economy operating in a balanced market, but a producer's system overrunning society while undermining human sovereignty, politically and socially, as well as economically. We need to remind ourselves again that the heart of the system, the corporation, operates only upon one purpose, the expansion of profits. That autonomous force is incapable of understanding or responding to goals of the society it effectively economically manages. Moreover, by the inherent terms of competition between autonomous corporations, there is no rational way in which the corporate economy itself can incorporate broad social goals or set reasonable limits into their operations. There is simply no alternative; the forces of competition compel maximum corporate aggressiveness, if nothing more, even for business survival. No limits can arise until the sovereignty of the people and their legislative powers can extract themselves from willful economics. So we are all tied to the corporate lemmings racing toward the raging sea cliffs of disaster.

Now that the industrial revolution has shifted society from the necessity of nature to the necessity of economics, we must ask new, powerful, and absolutely vital questions: Will the necessities of economics continue to drive us compulsively to a paradoxical condition of social and economic self-destruction? Can society find the means to lift the iron curtain imposed by competition?

A tragedy of development is occurring by an assumption of which endless growth can occur through the processes of a balanced market, but is forcefully expanded rather by endless demands generated by the arbitrary commands of compelling ever increasing and more irrational consumption. In the short term the process is waste, in the long-term disaster. How soon the disaster will occur and how it might be averted are the only debatable issues. The continuing growth of new "necessities" generated within the economic system merely highlights the absence of a broad social goal structure in the theory of democracy and a concept of human satisfactions to replace the insatiable demands of self-promoting, self-serving corporate economics. But today capitalism leaves that question to the consumer even while it promotes more frivolous and wasteful consumption. Such a system requires a new order that is ecologically valid, socially friendly, and culturally progressive. But the entire economic interest remains stuck on its very arbitrary aim, the unlimited growth of the GDP, as if indefinite growth of monetary wealth could fulfill not only important personal satisfactions as well as higher human aspirations. But, as it is, the corporate system effectively vetoes healthy evolution of the public interest, thus keeping all eyes on the sole purpose of expanding economic growth.

Yet, control of society is, strictly speaking, not the question. The issue is much greater. A society defined by economic terms of organization also strikes deeply into the course that society takes. As a consequence of this depth of corporate control, people are hardly able to understand and give clear human direction to public affairs. "People once accustomed to masters," Rousseau said in one of his aphorisms, "are not in a condition to do without them." What can be more conducive to shaping the structure of people's lives than an avalanche of technical change generating a profusion of goods? Valid thought focused upon human affairs is no match against corporate economic power and is made irrelevant by the covetous technical and economic advancement.

What society cherishes it organizes thoroughly; and in the opulence we live by today, it organizes with great panache. Unfortunately, what is neither cherished nor well organized is the critical course of society leading to the inherent well-being of the people. That must be our guiding light, not promoting raw growth, if we are to avoid the consumptive defeat of society. We must moderate indecent consumption and set new, more human visions of society.

Consumptive Economics

The source of our currently growing social and ecological disaster is that the system of corporate capitalism, upon which past economic development has depended, now relies upon a process of expansive waste and destruction. That disaster is as certain as the dependence upon competitive-driven process always seeking higher profits. The blind pursuit compels continuous growth for ever-greater profits. The objective of improving the qualities of human life is then lost in the pursuit.

By these blind forces of competition, corporations must exploit every resource and take every advantage to propel production and consumption to new, higher planes. The assumption is that consumption is proof of worth. But in reality the higher levels of consumption mean merely more consumptive waste. The higher costs of resources, the greater consumption, and the increased waste all become part of the game of raising company profits. Over time, as industrial production became more efficient, the environment into which the products operated effectively reduced the efficiency of their use, because more ecologically destructive, and demanded follow-up programs are required to overcome pollution and other debilitations, such as ameliorating polluted environments.

What is abundantly clear is that the system devours the very foundations upon which it stands and debases the human purposes it serves. In private enterprise, we are learning that the dangerous period of growth is when adequate living expands into affluence. Higher human attainments, hardly increasing, have new obstacles of errant material "wealth" to distract and beset them. And so a higher range of cultural achievements is easily lost to demands for greater output and consumption. Yet each step along the way of the induced process increases the overhead upon the people, government, and the environment, as when a freeway extension increases taxes, requires more government, and deteriorates the neighborhoods it cuts through.

The classic example of efficient production demanding inefficiency in its use is the automobile in the city; indeed, urban environments are now built for both automotive and urban inefficiency. That inefficiency is founded upon the exploded distances and time of urban travel, the approximately three parking spaces each auto requires throughout the city, the special construction of boulevards and freeways; and the great costs of vehicle repair, traffic police, signals and traffic controls, insurance and accidents, and school busing. *Wherever there is traffic there is a devastated environment for people*, as anyone can measure the impacts of traffic in various environments. And in the exploded environments of cities built to the scale of the automobile, the telling issue is how distances and travel time reduce participation in studios, galleries, theaters and clubs of urban life. Possibly the saddest features of all are the activities and facilities that do not exist because the urban overhead is too great to start or sustain them—a question we should ask ourselves when considering the estimated eight billion hours of lost time each year just in congestion delays.

Moreover, cities have grown as a consequence to an inhuman scale with important although unknown behavioral consequences. The blind promotion of automobility is an especially insistent metaphor of economic determination of modern development when the goals of society take leave of critical human values, as they have done in the struggle for unlimited growth of the GDP. That is the fate wherever the free enterprise system has captured the momentum of affluence with its enormous productive output, especially when it can operate on both sides of the open market through the urban structure and intensive advertising.

The issue requires taking hold of economic wealth to serve the social side of society, building true wealth to create penetrating human opportunities and leisure. And, if a new moderating ideology should be appropriate, it would give new meaning and purpose to the public-interest dimensions of life. We can see, for example, that if the form of the city is now organized as a structure escalating consumptive inefficiency, it can potentially be designed not only for efficiency, but also simultaneously for many forms of amenity and beauty that can be derived when the city is constructed at the human scale. That achievement can result in a huge efficiency comparable to today's massive industrial output, thus affording the leisure behind civilized accomplishment.

Restructuring Wealth

If, as it appears, society too easily assumes that what is good should be sought without limit. Certainly that is our pursuit today, like houses sprawled to consume the countryside, or more cars and freeways for more destructive movement. This fallacy of the infinite good is a lesson that applies to the entire economy, the incredible thought that infinity is possible. How could it be that the economy can grow without some form of an end perspective based on a vision of the human ideal? But now society needs to confront the runaway economy and its endless self-destructive growth.

That growing economy, even ignoring its destructive paradox, being sought as the supreme good, unlimited good becomes its own self-developed product, and leads nowhere except dangerously up for its inevitable plunge. That is today's pursuit, not merely of corporations, but of government itself and the entire society. The belief system behind capitalism remains solid, especially from winning the Cold War and in the trust given to the market mechanism. While there are growing doubts, there is little in-depth insight, and certainly no doctrines, to establish a rational basis to act positively for social development. Although ecological destruction is now increasingly understood, the current perception of issues does not lead to an understanding of the basic problem within the economic system itself. The environmental movement teaches us to recycle, but that approach distracts people's attention from grasping the more basic economic issues governing the environment that arise from the last century's most radical economic expansion. The prevailing belief remains that the economy is dynamic, responsive, and lies outside the need of fundamental reorganization. That such a system could lead to disruptive consequences not amenable within the market itself must lead to a new kind of public dialogue our society is not accustomed to. Any conscious redirection of the economy remains mostly foreign to both popular and academic thought.

A companion perception hardly in the public mind is that cultural enterprises should exist largely on the dole. That the finer possibilities of life might be set into the basic structure of society remains a flight of fancy, as witnessed in the declining fortunes of the national Institutes of the Arts and Humanities. But if the great sources of wealth now available but also seriously wasted in the runaway economy, are to be realigned for their human social abundance, the arts and humanities and many other such interests need to be considered as basic to the whole society. If we cherish the human side of society we should cement it firmly in the center of our organizational thinking.

For if, as I argue, the economic side of our civilized effort has been overwhelmingly achieved, the new basic elements of society are precisely those enterprises of human exuberance previously counted as luxurious cultural frills. They can become basic. The chief question becomes how we should best prepare and promote them. Since they are by their nature only slightly self-sufficient in their income, they cannot be counted as extensions of the private enterprise system. Something else, something more basic is required for a stage of social development that stands before us.

What is true of cultural progress is also true of creating urban community, an institution that might serve the full range of local human possibilities, including much of the arts and humanities, education and recreation. A constitution of community is required that sets it forth resolutely the purposes, environments, facilities, organizations, and—since the community reflects the many images and motives of humanity—the great variations community might take.

If there is one overarching issue now confronting society, it is that our condition of development has come to a radical condition and requires basic attention, both for social and ecological distress and for the vast array of human possibilities that lay ahead. But today the undiluted sway of money grips thought and the social habits of intellect are not ready to strike out into new territory, however critical the issues are, and however rich the potential may be. But we are ill prepared as a society to grasp beyond this watershed. Can society as a whole, therefore, learn new pathways to follow?

Economic Determination

Among the paradoxes of modern life are events of an underlying human discord. Cities are constructed to be wasteful and promote massive continuous consumption reflecting how advertising grew thirty-nine times between 1950 and 2000. The huge and articulate initiatives of the autonomous and determined corporations have implanted themselves as the definitive institution of society. These facts reflect one of the most profound events of the last century, that is, the rule of society by economic determination, an oppressive outcome we did not consider while endlessly promoted economic growth was universally supported.

A life-confining power of economics today has become basic to the radical transformation now exhibited by society: from hand to powered machinery, from workshop to factory, from local wagon transport to a national railroad network, from rural to metropolitan structure, from virtual subsistence to complete monetization of income. And in our day, economic determination is represented by the imperatives of automobility, jet travel, radio and television, computers and the Internet. The demanding economic grip dominates us through commercialization of sports and entertainment, the national unity of production and consumption, the proprietary nature of health care, the career grasp upon higher education, the co-opting of science, the penetration of politics and the purposes of government administration, the shift of women from the kitchen to paid employment, and now privatizing prisons and schools. Economic determination means the control of life and the perception of the human career as money, factored by one's personal cash register and governed by a budget always to be increased by every possible talent and stratagem. The agent of it all is, of course, the corporation.

Without exaggeration we can say that the rise of economic determination of social form, personal behavior, and the dynamics of thought are major results of the Industrial Revolution and today the no-less dominating Commercial Revolution, the first a control of labor to promote growth by industry, the second a control of the consumer to promote growth through commerce. We do take pride in the material abundance and the vaunted standard of living. But, by and large, we ignore the corollary effects upon our lives, many very positive and many quite negative, perhaps the force of economic determination being the blind end effect of a new necessity we have yet to appraise, let alone ameliorate or redirect. Urban sprawl and its counterpart in incessant urban travel together tells us much about being economically determined in the name of wealth, that is, wealth that in the end paradoxically denies us much personal freedom, wealth that is destructive of a sustainable ecology. In other words, wealth that reveals consumption as the objective of a dominating development process out of control, an infinite growth process without any clear human or natural end objective or value.

So the human transformation of the last century has been radical, a radical abundance on one side and a radical, if surprising, enclosure of behavior on the other. Social behavior, being encompassed by and limited to the strict terms of economics, is now swamped and subordinated to the controlling economic power. Economics took precedent when a family moved to the city and took wage work and became absolutely dependent upon cash income, that is, when the self-sufficiency of home-grown and preserved foods, home-made clothes, furniture and toys was replaced totally by manufactured goods. What was organic to the rural community—the local general store, flour mill, brick yard, tannery, and blacksmith shops—gave way all too suddenly, it seems, to large specialized plants, packaged store goods, extensive transport, and advertising. Consumption grew and offered new comforts, conveniences, and novelties, and promptly made them obsolete, increasingly centered upon throwaway paper and plastics. But early on, the growing economic determinism was understood as simply the growing standard of living.

As a consequence, economic influence grew steadily, and became more deterministic roughly each decade after 1900, intimately penetrating the span of human behaviors, taking time out only for depression and war. So the deterministic forces of economics found allies in the pull of new suburbia and push of the old central-city developments, and in the growing size of the city, the class-prestige trends of the times, and governments' readiness to encourage isolating urban environments by constructing the boulevards of commerce and freeways that expanded urban distances. All this was abetted by the real estate market, the American tradition of building and moving on, the mood of progress by dollars, and especially the new products and new wealth to obtain them. The achievement was a dizzying form-lessness of the city, yet it was the power that shaped modern living. All concepts of living, it seems, reflected the economic terms of urban living. Limits were set only by what could be afforded. The city, founded by economic determination, was not designed by social ideals or cultural aspirations. Those aspirations were confined by and even designed to be outcomes of economics, and hence were privileges scaled by wealth. Economics established the purpose of urban life, set the contents and put budgetary limits on cultural interests, not so much by absolute budget restraints as by being its own self-developing product and the dominating purpose, aptly noted again as the standard of living. So each decade expanded the economics of human behavior embedded in organizing life by production and consumption.

The significance of the industrial, commercial, and urban revolutions therefore reduced itself—in deep human terms—to the economic forces of behavior while erasing physical and social patterns of association existing for a millennium. The basic power was corporate and the sole basis of corporate action was profit. People were organized *by* their purse, not *for* their freedom.

Corporate Control

The corporation's pursuit of but one purpose of profit is fundamental, permitting it single mindedly to build its strategy of investments, exploit natural resources, develop technology, manage labor, and shape behavior to market its largesse. From that single profit motive came the stream of new and better products at lower prices for more buyers with higher wages capable of purchasing along the phenomenal range of goods and services. That was the unique revolution of wealth that is unlikely to see another century like the twentieth.

By working on both sides of the market with its sophisticated marketing, and by the normal economic process of refashioning the growing urban population into masses to be processed as manageable workers on one side and readers, listeners, viewers, and buyers on the other side. And by this processing of the splintered masses, the corporations could inhibit or deny other distractive social interests of people that might interrupt them from being maximum possible consumers. How effective this process has been, one needs merely to consider the non-evolution of social experiments in the twentieth century, something more conscious than the hippie non-consumer movement of the 1960s. By playing upon the entire society and controlling the major media the corporations could pursue their profits within what Henry Steele Commager called "the chaotic creedlessness of the twentieth century."

As one may discern, the corporations, which operate smoothly with the many processes under their control, create a basic social chaos for people outside of the strict market process itself. While individuals who could earn a steady and reliable income could also participate as a consumer with a seeming freedom and understanding, in their social lives they are left open, fragmented, and without meaningful social and personal coherence. Churches, clubs, and other memberships, while serving special purposes, also become merely elements of a social laissez faire as fragmented as the economy, for each person to confront individually as he can. It was not difficult, therefore, for a company to relocate an employee from Houston to New York or from Boston to San Francisco. Ties to the employee's home city are usually not significant, perhaps little more than in-laws or a health club; nor are they associated meaningfully to co-workers, which in any case is an alienated behavioral setting. And, for frequent travelers, many persons become better acquainted with people in cities across the country by visits and telephone than they are within one of their neighbors but minutes from their door.

These social fragmentations were underscored when after 1945 the diminishing abilities of downtowns to hold onto commerce meant that one must go in numerous directions to malls and commercial centers to obtain desired goods and entertainment. One planner has termed this growth of urban functionalism as exponential hardships, a penetrating life-long, social handicap, which is a comment on human inefficiency that has hardly surfaced in public dialogue. Such interpersonal chaos is but an extension of the corporations' ability to define their power. Even the corporations may not be aware of this item in their quivers of power.

A mere listing of the corporate powers of social behavior reveals their daunting nature: (a) the exploitation of the nation's airwaves without charge; (b) the intensity of advertising that penetrates the social mind; (c) the co-opting of traditional sports, music, and drama as front material for advertising (d) the real estate formation of systematic waste structured into urban form; (e) the management of bio-engineering; (f) the proprietary control of health care by pharmaceutical, insurance, and HMO corporations; (g) the channeling of all levels of education for confined corporate economic purposes; (h) the corporate welfare received from government; (i) the huge political contributions to legislative candidates and the intensive lobbying for favorable legislation; (j) the growth and centralizing of strategic corporate power resulting from mergers; (k) the growing grasp of global power illustrated by the emergence of the World Trade

Organization, giving important powers of international governance over nation states.

Especially in combination, such varied and penetrating powers over the society and behavior of individuals is historically unprecedented. Governments in peacetime cannot ordinarily compare to the collective corporate forces: neither in independent action nor the freedom of initiative; neither the diverse fields of action nor the varied strategies and tactics; neither the singularity of purpose nor the sophisticated technologies and bureaucracies; neither the penetration of local politics nor the impact upon national affairs; neither the rights protected by the First and Fourteenth amendments nor the control of barren, incessantly mobilized urban environments. Thus the smoothly functioning power of corporate economics—so positive, yet so subversive—is sharply contrasted to the human chaos of the pervasive market-style organization of society. The primary basis for personal adjustments, then, is to find happiness locked into the motivations for wealth and higher consumption.

We need to be absolutely clear about the human significance of corporate power, which rests in its unique ability to organize all possible resources to be processed, first to propel output, and second, especially after 1950, to propel consumption—that is, the combined impacts of the industrial and commercial revolutions. That vast processing is close to the fundamental nature of corporations, whether it is iron ore to be processed into steel, milk to be cheese, or land to be occupied by houses. Money is the fundamental processing agent and the basis for all other economic processing. As the resources have expanded to large-scale employment and mass marketing, people themselves inevitably become resources to be processed, no less than cotton or coal. Private wealth, seemingly a power of freedom, is the special target for promotional processing because freely expendable money and wider margins it offers for advertising and sales. Quite monolithically, the totality of economic processing exposes the entire society to the combined corporate manipulative powers, effectively creating a new form of monopoly—over the entire society—as a supreme corporate province.

The economic processing is essentially faceless, reflecting the intentional alienability of money, faceless social functioning, and a faceless power, indeed being a non-human, alienating ideal form. To be faceless—to be alone in the crowd—is to lose the most human essence of life in society. But today facelessness serves the ideals of material, money, and power, and does so to the core of our social being.

It is the faceless system, therefore, that is becoming the final result of the industrial, commercial, and urban revolutions—all manipulated by and for wealth-making. Then materialism easily becomes more of a bribe than a true human benefit. Social and ecological crises are then inherent because the system knows no limit, confines aspirations to market processing and denies public avenues of progress, striving instead to endlessly exploit all possible resources and having no responsibility beyond profit.

When we learn to measure the scope of the runaway economy's fundamental role of destruction, we will then be able to view people and their earthly habitat as the superior values and economics as the subordinate instrument.

As we continue to automate output, the no-hand output can possibly relieve people from systematic oppressive processing. This opportunity will permit us to personify life as an inspiring objective. But, as the economic determination so powerfully demonstrates, it will not arise from any "free-market" interest or corporate intent. The very nature of modern society demonstrates—especially in the socially barren and ecologically destructive cities—society as a whole must take a grip to make social change creative. We must think through the human possibility, and set a new course for a very different library of human accomplishment.

Chapter VII

Democratic Imperative

Our concern here is to develop a perspective in which the monolithic growth of the economy will give way to a central focus upon cities and the prospects for an urban democracy to elevate the sovereignty of the person to its rightful supreme position in society. Sooner or later the overwhelming force of economics and power of corporations must be moderated and a new focus on the life of people in cities needs to be proclaimed and given directive force in society. While our era of economics was brilliant and successfully brought affluence to the majority of Americans, that era stressing economics is essentially an historic sidetrack, which is by its own terms a supporting service to society.

Organized Corporate Power

> *There is no society in which everyone does every-*
> *thing for love; but a society in which no one will*
> *do anything except for money is in serious trouble.*
>
> Francis L. K. Hsu

> *Freedom requires the overcoming of alien-*
> *ation... The principle source of today's alienation*
> *is not a ruling class, but a social process domi-*
> *nated by bureaucratic institutions.*
>
> Paul Goodman

Society today is the first to be built on comprehensive power. That power is technical, dominating nature. It is economic, now organizing in one scheme the resources and work society. And now it penetrates the social fabric to become the most complex and awesome power of all, for it reaches deeply into the formation of personality, popular behavior, and the definition of cultural goals.

The emergence of power to this scope and penetration, at once both unified and defuse, is the social phenomenon par excellence of the twentieth century. The most comparable event in history, perhaps, was the slow emergence of agriculture, where people domesticated themselves while domesticating animals and crops. That process consequently propelled the formation of central and continuous political control by dynasties of pharaohs, emperors, and kings in a blind process of change. And so it is today. But this time the stark new powers of domestication have arisen almost within a lifetime.

Society in recent centuries limits state power through constitutional democracy. Limits on industrial power have evolved that forbid monopoly within an industry and regulate labor practices, food and drugs. Now we are beginning to moderate adverse environmental impacts of industries and their products.

And when we look at the roots of social power, the greatest initiative and force nevertheless exists in corporations. That power combines massed capital, the most advanced technologies, sophisticated teams of varied specialists, integrated systems of production and especially today, the promotional power of marketing. A new model of dominant corporation is appearing, one that is more commercial than industrial and focused more on the uses of wealth than basic human needs, and grows with each new level of freely expendable wealth. Each increment of family income enlarges the scope and profitability of companies that can influence tastes and shape behavior as never before. Corporate power thus shifts while it grows.

The classic industrial giants, like Carnagie, Rockefeller, and Ford, brought society into affluence, fulfilling family necessities and advancing the range of conveniences. But later, with the appearance of greater affluence, emphasis shifted to luxuries, travel, and other choices. Accordingly, corporate marketing strategy expanded and shifted significantly from products to the consumer, far less upon products or services as such. People have more freely expendable income that may be spent in many more ways, by sharp contrast to the time when bare necessities completely dominated family budgets. So companies now work to influence consumers as much as serve them. Where quality and price once dominated promotion, emphasis has shifted to create an idealized image of life associated with a product or service, be it whiskey, tours, adventures, entertainment, or luxurious automobiles. Direct material benefit has thus slipped in relative importance. New corporate power resides in this strategic shift and seeks to perfect its ability to influence public tastes, if not entire life-styles.

Much has yet to be learned, of course, about consumer-penetrating powers of corporations in the new conditions of wealth. Quite plainly, however, a major basis to support this new sophisticated commercialism is found in the physical form of the city. We have seen how ill defined cities create an almost endless demand for product, especially in the escape to costly suburbs requiring the public to construct boulevard and freeway networks to serve the unending urban travel. What is most telling about this planned chaos is that the escape to presti-gious but socially barren suburbs and the accompanying frenzy of movement also establishes both the physical circumstances and the psychological conditioning for the commercial influence of people's social behavior, leisure time, and cultural ambitions.

By an astonishing coincidence, the rapid adoption of television occurred after 1950 appeared precisely when affluence first became widespread and the rush to the suburbs and the construction of urban freeways established their consump-tion-demanding patterns. Families moving to far suburbia condemned their chil-dren to an appalling social isolation, even more than to themselves. This exposed the children to the new intensity of advertisements made possible by television and to the purposely low level of programs demanded by the advertisers. Here, then, was an unprecedented opportunity to exploit the combination of new domestic wealth, new isolation, and the addictive novelty of television. Here was a vast opportunity and broad challenge to implant the commercial visions of the good life deeply throughout affluent society.

The objective of promoting ever-growing mass consumption as the ideal of life was anchored most deeply in the two most consumptive urban products, the spacious house and a second or third luxurious automobile. But the house and car did more. They opened two additional lines of mass consumption. The first was a great range of equipment required for the larger house and yard, notably yard machinery and the progression from high-fidelity records of the 1950's to the panoply of today's home electronics. The second included boats, skis, hang glid-ers, camping trailers, motor homes, snowmobiles, dune buggies, and other equip-ment that could be carried or pulled by car and stored in the garage or close to the house.

The changing form of the city therefore provided both the organizing founda-tion and psychic force behind the ideal of consumption. While urban form underwrote the promotions to make consumption an all-encompassing ideal, it did far more. The *enforced* mass consumption onto families to their maximum ability. While the house and yard required or facilitated endless accumulations of recreation equipment, the spread city demanded not only multiple family cars, the isolation also made necessary the in-house entertainment of television, which could then directly promote consumptive visions of life. Thus the metropolis became the great silent partner promoting ever more massive materialism.

We first build dysfunctional cities. Families are then trapped into a new scale of more prodigious production and consumption. People then fall more deeply into corporate promotional programming of corporate television entertainment, managed news, orchestrated broadcast sports, staged quiz programs, directed tours. Both the physical circumstances and the psychological persuasions then deepened pecuniary values and elevated material wealth into life's greatest goal.

Commercial power thus appears by four mutually reinforcing factors: (a) the urban form physically demanding a high level of consumption, (b) the large scale of the house and several cars further accommodates and induces massive material consumption for leisure and recreation, (c) the socially barren circumstances leave little alternative to the offerings of mass consumption, and finally (d) numerous powerful and persistent advertisements and media programs create a value system that emphasizes how life is less than worthy without great expenditure and material luxury.

How has society become socially so intricately locked into the power of economics, so thoroughly subordinated to commercial power, and simultaneously so nearly blind to the many resulting social discontents? Certainly the social, political, and economic dynamics have yet to be described, let alone put on the agenda of democracy. The overwhelming national concern by government for economics makes little sense when so many other public crises and opportunities are given but passing notice—yet also create critical financial pressures. The whole of society, it seems, is under the spell of the siren song of an historically unprecedented and unfamiliar wealth—naked, socially undigested wealth, at once supremely addictive, heavily promoted, and surprisingly self-defeating.

Material wealth beyond necessity, beyond comfort, is valid when applied to the many larger human dreams. But the wealth we have today is too forcibly built as newly necessary consumption, too much a part of waste, too dangerously a part of popular manipulation, and all too much a source of social crises. That economic power—power created so naturally in the course of doing business—can penetrate and even define the central essence of modern life and has become a travesty upon the human progress and a woeful nightmare of democracy.

Modern economic power raises basic social questions that are radically different from the old Marxian class struggle, a struggle founded on an extremely limited wealth. Here, rather, we see a most promising growth of great social wealth going amuck. Today we see a new power controlling people in the name of wealth, a national fixation upon an extraordinary narrow vision of economic well being, all against a backdrop that covers over the mutual destructiveness occurring in cities and social life.

What kind of entertainment typifies that shaped by the corporation? Each program is planned for a charge, as in cinema, or shaped for advertisers, as in television. Television must fare well with the Nielson ratings, a measured sounding board for effective marketing. Public controversy concerning programs must be avoided, therefore, and the programs must never go over the heads of viewers. Programs are carefully staged for an extravaganza effect (music, theater, sports, or quiz) that avoids serious public questions and dialogue, and censors out divergent views or entire issues. Only then do the honey-coated programs become effective settings for advertising.

The astonishing powers that are increasingly evident today are built upon a profound blinding process. That process is varied, comprehensive, and mutually reinforcing, and self-propelling. It carries its own logic and is built deeply in American traditions: the land of opportunity, personal freedom and independence, the private enterprise system, the sanctity of property. The blinding process itself is founded upon: (a) the physical circumstances endlessly demanding goods and increasing waste; (b) a promotional process creating an ideal of consumption; and (c) an older, deeper ideology equating money and materialism with personal progress and national destiny.

Here is a legacy of the Marxian pall that severely undermined social thought throughout the Cold War, an oppressive thought control we must yet overcome. The work of the conservation movement is the best example of a counterforce moderating corporate power, although that realm is but a small slice of the more comprehensive issue. Still closer to the broader question is the entire form of the urban environment.

Disorganized Loss of Personal Power

> *The lack of creative response to [urban decadence and ugliness] is particularly discouraging because all thinking persons are aware of the situation and are anxious to do something to correct it. But common action cannot be mustered because it demands a common faith that does not exist.*
>
> Rene Dubos

To the extent that today's city is broken down into an extremely low common denominator—individual parcels to be exchanged as freely as any product, the fracturing imprints itself into the mind of persons, into the form of public affairs, and almost universally emphasizes dollar values. If the mind perceives the city as but an aggregate of parcels infinitely alienable, then that mind will have difficulty imagining any higher urban purpose than that exhibited by the collectivity of urban particles. It is indeed ironic, therefore, that the largest and most unified programs of cities since 1950 are the construction of boulevard and freeway networks that give ready access to the subdivision of lands into millions of new fractious particles. Here the pecuniary values associated with endless parcels come face to face with the threat of a complete breakdown of urban form and function. The struggle to construct highways that tenuously hold the city together then becomes the central and dominating struggle of thought and action about cities.

When the city is thus pulverized into endless exchange and the physically forced expansion of movement, the person's psyche also becomes pulverized into all of the bits of activities required to function effectively in society. People effectively tend to become as separable, as exchangeable, and as loose as the parcels and products they deal in. Commercially and psychologically, the dispersed city and the massed population are both reduced to their lowest common denominators, that is, to small particles of land and isolated individuals that can be dealt with as a total mass through one great process of production and consumption.

Thus, to the extent that people are thrashed to their smallest exchangeable bits—the individual and family—human relations are reduced to their random, shallower, pecuniary values just as forcefully. As deep human bonds diminish in the process (even within the family itself), the demanding logic of the dollar converts all values—and people—into one common commercial market of exchange. In this sense, enterprise thrives on social fragmentation, because people are then open to commercial exploitation. Where enterprise has free and unfettered access to individuals in a mass market, to people whom otherwise have less stable social roots and who are seriously alienated, their attention can therefore be psychologically monopolized by the incessant commercial appeals.

Partly as a consequence of land being fractured into small sizes for maximum exchange, the availability of public spaces open to all for free interaction has declined decisively. Where once downtown areas were common spaces for free casual association and festivities, today's dominating shopping malls are now private and lock their doors at set times. One is welcome only as long as a sale is in prospect. Social *loitering* is only slightly tolerated by those that now own what was once the most social of spaces in the very public downtown areas.

Where can people go, then, to be social and urbane when downtown streets are no longer social or urbane, and attractive, when common and informal public gathering spaces decline or have effectively disappeared? People can, of course, go to the restaurant and theater, perhaps a museum or gallery. But here again, they are institutionalized in the way the institutions want to operate. With the critical relationship being between the event or exhibit and the person, interpersonal association becomes socially incidental. People are, very simply, limited to the terms organized by the great economic monopoly.

As public spaces decline in favor of the shopping malls—the money machines—the only places where *public* spaces have effectively increased are the freeways, boulevards, and parking spaces. These have no social value, but are, again, money machines for auto manufacturers, oil companies, trucking, and road construction. Can people awaken to these infringements of rights we never knew we had? Can people awaken to the increasing manipulations and spatial confinements restricting the simplest, most casual forms of human association?

If the environments we make mold us, the cities of today are formed as massive accumulations of human particles that are largely unrelated except through the machinery of economics. People become like unorganized laborers that once had to face the manufacturing companies individually. Equality of bargaining could not exist until labor was able to unite and speak as one.

So it is with today's form of cities. Traditional social bonds have been pulverized by concentrated industrialization, aggressive commercialism, and mass urbanization. People are unable to define their own lives. But unlike factory workers who once singly confronted industrial power, people today find no single power over their lives to blame. Such miscellany structured into behavior, while economic at its source, denies or disrupts the social essence of personal behavior. Corporations do not need to organize to create their monopoly.

Consequently we face a fundamental defeat of democracy. While basic democratic institutions seemingly remain sound and wealth generally abounds, people are unable to build credible lives outside of the terms of their economic performance. They cannot shape congenial and socially efficient environments, and they lack the organizations that could reestablish meaningful social bonds. More critically, there are no concepts to even debate the nature of social progress outside economics. Since there is not an understanding of what creative interpersonal opportunities in the modern context might do for people, the idea of a high standard of living is not a social concept but nevertheless dominates social behavior. So corporate definitions of the good life maintain their cultural monopoly over the mind.

Naturally, then, the overwhelming effects of economics throughout society cause us to lose sight of the many grievous human crises that boil out of the cracks of the misshaped and bloated economy. When we create physical and social environments without endearing qualities and stabilizing forms, and substitute market exchange mechanisms for them, we need not wonder long why we face constantly growing social disintegration. The people who are expendable at the bottom of the economic ladder are the greatest losers, for they suffer the fragmentation of life and are bombarded with the persuasions intended for the affluent, but cannot share even the narrowest economic vision of the good life. That fact alone is sufficient cause for a clutch of social crises.

When we have eradicated so much of the naturally occurring social-immune systems found with strong, stable families and communities, we force society to inordinately expand its professional and institutional services, the unnatural corrective activities for social disabilities promoted by organized society: family and child protective services, health and psychological assistance, teen pregnancy services, special street police and drug enforcement, expanded courts and prisons, corrective education.

Here, then, we see two dominant effects of economic growth and two classes of citizens. The first creates a society of individuals whose vision of life is confined to the mercantile ideal of output and consumption, competition, and materialism. These people are the winners. Their behavior is motivated around higher expenditures, that is, buying a standard of living rather than building socially creative life styles.

The second kind of economic effect just as firmly creates a society of socially disadvantaged people who, in almost an inherited way, carry their disadvantages from generation to generation, seeing society as illogical, unfair, and threatening. So they act in kind and perpetuate the system, despite the social expenditures that fail to alter the intergenerational defeats.

In the pastoral, pre-industrial life, economic organization was an entirely secondary matter in society, that is, economics was submerged in social affairs. That life also coincided with a comfortable scale of interpersonal behavior, whatever the material limitations or political confinements there might have been. However, with the rise of mass industry, mass mobile labor, and undifferentiated mass urbanism, the economic and the social organization of society diverged sharply. While economic activities exploded in scale, the old rural and village social order never found a comparable or congenial scale in the new massive cities. Lasting, meaningful, social organization only fractionally formed there, losing out to the struggle for material progress.

Thus, while corporations succeeded in freely amassing and organizing resources, mobilizing labor, and building distribution systems, a basic dissociation occurred with the masses of people in the vast amorphous expanses of the city. This loss is expressed in the majority of neighborhoods that typically exist today without social focus, boundary, organization, or any significant social unity, tradition, loyalty, or vitality.

When people are effectively fragmented, forced to follow employment anywhere, they are taught to forego social values so that they can purchase the benefits of life. When they live under conditions without regular, rich, and comforting association, their lives necessarily reflect the new pervasive money values of the economically dominated society.

Still, corporations find no democratic limits confining their radical reshaping of cities and social organization of society. Their singular mandate to earn profits by all available means thus effectively controls urban form—the whole living and natural environment. Public planning controls do not really shape the form of cities; they do little more than establish standards of physical development, tidy the process, and assure highway requirements. The physical city in turn effectively limits the kind and character of social organization. This means that the social organizations that do exist, the associations, museums, galleries, and such, remain few in number, dispersed, limited in function, and largely incidental to urban life.

Apparently, we have accepted the proposition that capital is the preeminent, legitimate force governing the form and texture of cities and society, that people on their own initiative are minimally operative beyond consumer choices. Consequently, their power to shape their own lives is pulverized by the fragmented urban form.

We must not be drawn to the belief that attending church across town or a Tuesday Lions club are socially substantive. They are only bits of meaningful social organization, good in themselves but lost in the sea of a society otherwise organized. Building friendships has become a logistical hardship. A common ground hardly exists for the casual, regular, and mutually free interests of people to evolve naturally.

We still live in an early history of the industrial-made society, when the person is largely powerless except for purchasable benefits. This condition has been disguised by the rise of affluence, the marvels of new products and their manipulative inducements, and the feeling of control over one's own expenditures. But the very structure in which expenditures are made makes a farce of the apparent freedom of expenditure. What matter is it that the choice is a Ford or Honda when one is forced to purchase another automobile?

Our time will be known as the age of materialism. Maybe in time we can learn how naked material power really is without deep human aspirations. Perhaps materialism and the weighty domination it has over our whole existence is but a prelude to different vast personal, interpersonal, and social possibilities of life.

The blunt fact of the corporate power and pulverized personal conditions stands starkly before us in the great economic monopoly over the whole of life. Someday we will learn that our lives and our environments are too vital to be managed only for exchange values. But for now, the corporate goals and corporate initiatives have become superior to a social organization of society. The overriding question of human power is to find a meaningful organizational basis for people to empower themselves outside of the rigidly confined, economically directed consumerism. Production and consumption are both prisons if the circumstances of life are organized for little else; both now represent rights we did not know we had. Hence they constitute an insolvency of democracy.

If people are to effectively empower themselves, they need to develop environments and organizations that intimately reflect their daily and long-term aspirations—indeed, that also stimulate such aspirations—while simultaneously establishing an effective *competition* to the pervasive power of profit-seeking and profit-limited corporations. Such conditions of living, I believe, should be local, at the human scale, and promote a wide span of participatory environments, associations, and activities. That is, they should create a broad range of personal opportunities.

Today our condition is organized power for corporations and disorganized powerlessness for persons. The imperative of our time is to establish a human primacy in the organization of society. Above all, this means the social and physical form of the city.

Socioeconomic Monopoly

> *Modern confusion and distress now reveal themselves as a consequence, not of human evil, but of the process which provides material abundance. That process has assumed a vitality and form of its own....its necessities of ideology, value and organization impose themselves on the social order...the apparatus of abundance have so fused with our environment, life styles, values and thought that it is no longer possible to know who is the instrument of the other's purpose.*
>
> Paul Goodman

Our age is like no other. Our powers, problems, and possibilities are like no other. There are, therefore, few lessons from the past to call upon for guidance. We confront conditions today that suggest unlimited, almost absolute powers, especially in various combinations of high technology, huge corporate bureaucracies, and immense capitalization.

In addition, we see in contemporary society the appearance of a new phenomenon, the *collective* powers of huge corporations. Here is a new kind of power that ranges through mass media, advertisers from many industries, most urban development, and all manufacturing and services that benefit from urban dispersion, disunity, inefficiencies, chaos, and social disturbances. These forces operate together with a fundamental unity of action. Whereas the era of rapid industrialization was concentrated on physical product, our affluent era of commercial development is characterized by a corporate concentration upon people for corporate purposes. Society does not understand this power. Consequently, there is much room for abuse, especially when the character and content of abuse has grown by many degrees, appears normal, and is not socially or legally brought to the greater human purposes. But serious abuse there is, as certainly as we have seen in shaping cities by commercial exploitation.

The unprecedented form of power today is very real and present, pervasive and active. It grows upon the organization of numerous resources: all raw materials, most land, all tools of mass media, virtually the span of technical skills, and great ranges of professional knowledge. This great force is profound, sophisticated, and directed by special interests only for those interests. Affluence is the key to that power. The city is a primary field of operations.

Society thus confronts an ominous economic monopoly.

This completely new power is a true monopoly because, however diverse the sources and methods of its power, this monopoly subdues every person as a person, and is unopposed, for there is no competitor, only a common competitive zeal. There are no limits to its penetration into society, no confinements to its domination of the popular mind, and no limits to its defining and confining social purposes of our society in our age. This self-aggrandizing power is not democratic, and it threatens representative government through massive funding of candidates for office, intensive lobbying, and directing the mass media.

There is little resemblance between this new power and the classic monopolistic domination of one industry by one company. Government does not affect it, for there is no culpable abuse under present law. Since government remains focused on traditional questions of competition and regulation, society has little to say about the penetration of this power into all social life.

The economic power of monopoly is evident in two ways. The first is the sheer concentration of capital, the highly organized strategies of action, the penetrating lobbying, and the general mindset that economic interests in society are paramount and should be directive. The second side exists in building an urban society that is massed, accessible, and manipulable for market penetrations, and thus that life has become foremost an economic affair.

In a society in which citizens have stable socio-cultural roots, clear humanistic aspirations, and strong supportive organizations to collectively express and promote those aspirations, the economic monopoly over social life cannot exist. But when the purposes and organizations of society are governed by thousands of organizations whose identical objective is defined by profit and managed through aggressive competition, a subordination of people for production and consumption—and for their prescribed way of life—is the natural and compelling result. That result, then, is the economic monopoly of life.

Here again, we face very new historic conditions in which useful precedents hardly exist. To our distress, the economic monopoly of society has become fundamental to our lives. These powers are diverse, persistent, and subtle where formal collusion among corporations is hardly relevant, where there is but scarce *competition* against commercial dominance in social affairs, and where there are complexities that disguise the corporate power and make it appear to be normal, fair, and progressive.

In our concern for the monopolizing power of corporations, it is critical that we focus on both the vital issues and high potential of cities. We must learn that the future of human progress no longer centers on raw economic growth, or productions and consumption, on blindly submitting to economic determination, on continuing to expand the runaway economy. We must be done with pursuing the endless good, in economics and especially in endlessly expanding urban automobility, which has unfortunately become the endless necessity.

One day we will give the economy the status we now bestow upon agriculture, applying modern technology as appropriate and governing economics like the demand for various foods directs agricultural production. Decades ago surplus food output was an agonizing national issue; now our agony is upon destructive runaway economic output, which we must also bring under public direction.

These factors that brought us to economic wealth are now destroying both wealth and the freedom they seemingly advanced. They put us at the doorstep of an endless urban functionalism, the urban ugliness and the multiple neurotic stresses. They have created a system of destructive waste and an economic and urban structure promoting the culture of crime. The singular focus on economics has given us an urban ugliness, a disarray of urban environments, and the deprivations of social abandonment. The aggressive economic expansion accompanies an anti-public ideology and further promotes dollar-dominated privatization, that is, the singular stress upon a false efficiency, now focused upon profit making of water, prisons, and schools.

Must we, can we terminate the domination of economics and stress the public interest embedded in cities? And can we center our attention on the whole of cities—physically, ecologically, socially, culturally, and especially economically—for that comprehensive concern will inevitably center upon the well being of the person. The purpose centers on the sovereignty of the people, and cities are the environments serving social achievement, if not human triumph.

Have we fallen into a trap like history's first agriculturists, who built the framework for horrendous centralized kingdoms to arise, which required ages for society to conceive and develop a democratic society. Are we not now building a dynasty of corporate management awaiting a future age to define a *socially democratic urban system*?

It is for all people and society, therefore, to understand the forces that do powerfully confine and direct our lives and then to build the environments and organizations that will strike a new balance strengthening democracy, especially a democracy very close to every person.

Organizing Imperative

> *Through the exercise of private liberty we are made to forfeit the possibility of association and intimacy which is the premise of individual power.*
>
> Paul Goodman

> *There is a deep hunger in our land and throughout the world for a new sense of community.*
>
> Agnes Meyer

In this era of the greatest human transformation, we point with deep pride to how in one lifetime people, who in their youth rode horse and buggy, could later cruise in air-conditioned automobiles at sixty five miles per hour and fly across the continent nine times as fast. But in the old towns, most people knew one another, or of one another, and this condition signified an organic quality of social relations, reinforced by the extended family and nearby relatives. The loss of that quality is what people lamented most when the population grew, industry appeared, and anonymity prevailed. Left to themselves in massed undifferentiated cities, people simply became deprived of age-old ways of associating.

Yet the profound transformation did not just happen. It occurred because the development of massive, complex, and varied products required equally massive, complex, and varied organizations, resources, and planning to achieve new, more phenomenal results nearly every decade. The transformation could not have occurred without a radically new condition of mind and a very specific and narrow canalized direction for creativity.

And in this age of technology, the demands of industry have received priority in nearly all of the ways we do things. Largely as a consequence, we have seen that the organizing genius of humankind has permitted the traditional and intimate side of its social nature to quietly dissipate in the face of the forceful and noisy forces of making things by ever larger industrial and commercial conglomerates. And thus, by organizing society to produce phenomenally, a society of quiet desperation has indeed appeared throughout the cities. More poignantly, that quiet desperation has been replaced in many neighborhoods by an explosive desperation when oppressive economic inopportunity also enters the social equation.

When we organize society for industry and commerce, we not only garner personality and creativity for that purpose, we substantially reorganize the very roots of society for that one monolithic objective. The subordination of the human side of society is much like that employed during World War II, when the war effort massively subordinated people. We understood and openly accepted such subordination. But now, in the much longer span and under normal conditions of change, we must certainly ask why people have been left in the dust of their own accomplishments.

The style of mind and organization present in our age has become radical, not only in its technical professions and high-tech prowess, or in the resulting huge concentrated wealth it has created, but in the broad social depression that in some conditions approaches depravity. We can say this because the unprecedented wealth masked the deeper social issues of how people live together and make life great or onerous for one another. Since, above all, we are deeply and inherently social beings, it is in this illusive but critical realm that the malady of our modernism permeates us all—and diminishes our humanity—to the core of our being.

While we were reorganizing today's society, we invested our full heart and soul in its products and luxuries. So our minds and our creativity are infused with the economics of life, it often seems, almost to the level of life itself. We organize the urban environment and our dominant institutions—even government—to build the mass, complexity, and power of industry and commerce. Nevertheless, it is astonishing to observe how most sectors of society, from citymaking to education and health, are so regularly interpreted in the terms of economics. A good city is seen as one that is growing economically, far more than improving itself as a place to live. The development of living areas is managed almost solely as a housing market, not as the creation of socially valid communities capable of responding to the diverse human needs or in expanding the social and cultural opportunities in healthy human settings. A good education is seen as one that will improve one's potential income, hardly at all one that is inherently good for the full range of living.

But the transformation of life is not over. The heart and soul of society can reassert itself and may be starting to do so. With conscious effort, a new, specifically *social* vitality can be reclaimed, renewed, and expanded. More important, the very wealth of industrial society can now underwrite an unprecedented *social* wealth. In a democracy, the only logical direction for development is the promotion of a supremacy of the person. All else is secondary and supportive, if it is not to be destructive.

Yet even as an aspiration for a new social dimension of development may now be appearing, we know that little worthwhile happens in life without its being precisely organized and set into the clear objectives and methods of society. Society encouraged original industrialization with large grants, subsidies, and protection. Thus, governments and corporations together built the railroads in the nineteenth century and then the highways and airports in the twentieth. Homestead and mining laws and the land-grant colleges established the framework for individuals and corporations to organize for the unknown but nearly limitless opportunities ahead. The government's laissez-faire approach to business was, then, itself a policy promoting powerful organizations that could combine resources and shape the markets for vast and varied expansions.

But no such grand perspectives, or goals, or supportive programs served the populations that migrated to the cities. Affordable housing eventually included inside plumbing, electricity, and varied appliances that defined the new standard and eventually ushered the first large populations into the age of affluence. Yet nowhere was there a public perception, let alone a system of public organization, that could build socially and culturally significant local environments and institutions for the urban population. Purely material and commercial terms of progress prevailed, and these were organized to benefit the commercial interests, and did so enormously. So today, we are paying the price of decayed inner cities and sterile suburbs, freeways and commercial boulevards, crime and drugs, and uncontrollable social disorder.

Today we face a suppressed social vision and lack the vital organization for an unquestionable *social* progress. While the structure of the city lacks organic human integrity and has proved to be dysfunctional, government under the corporate thumb clings fast to the rule of minimal governance, not optimal service. We do not build environments of positive social values. The failure is both conceptual and organizational. Americans have prided themselves on being a most practical people. Yet in urban terms, the more advanced our technology has become, the less practical our physical circumstances of urban life have become. And so we face a fundamental contradiction that will remain with us until an equally fundamental urban redirection can be initiated.

Until a modern renaissance can generate new directions of thought and action, the grave debilitation's will grow and public reactions will consist of bewildering rearguard defenses, like the blind reaction of putting more police on the streets, a sure sign of defunct economic thinking. We have to think big about cities and society. Cities are basic to the roots of our modern being, spanning most hours of our lives. They guide our feet in action, can direct our hands in art, our minds in phenomenal creativity. Society is encompassed by its cities, wherein we all hope, aspire, strive, and bring to fruition what we are as a people.

A renaissance can build cities, and cities a renaissance. Where else will we find the inspired source, the dynamic medium, and the grand creations of renaissance more closely united than in the cities we make?

Can we make cities the focus of popular inspiration, organization, energy? Can we make cities of aspiration, built to cradle the mightiest of being in every person? If we can, we will then learn to convert our technical and economic prowess into rich and diverse human rewards?

But to date, the organizing role of governments and the integrative role of cities have been denied to the public. Under market control, citymaking is accidental to land availability, organized for salability, aggressively marketed and stretched to everyone's ability to pay, which then stresses class and money values, separates the home from services and activities, promotes endless travel, and preempts government budgets. Therefore, we speak of housing markets, but by itself housing has no organic unity or vitality for a valid public life.

Society in the industrial age has now reached the stage where virtually all of life is defined by management processes. That is, the content and character of the whole society itself is defined by conscious terms, formal organization, deliberate action, and critical evaluation. That purposive power shapes even personality. But as yet we do not follow this logical process for social development. The time has come to apply the deliberate systems of science and administration, design and integrated development, and the ideals of democracy to cities, for they are the most fundamental physical and organizational fact of modern life.

In our modernity we know this: Only what is consciously organized and developed can exist, or assure positive results. Industry has taught us this new potential of organization. Can we now use that organizational power to make the whole society worthy of our progress?

Since this age has more to offer us as a people and as a democracy than any other, we cannot allow the wealth and machinery underwriting our prospects to divert us from those prospects, even temporarily. In time, an awakening is inevitable. Yet we cannot leave a bountiful future to chance evolution, as if the future were merely awaiting a string of predictive mutations or random inventions, as in the past.

Good cities are complex and coherent units, not fragmented accumulations. Therefore, a different mindset and a new collective will are involved. So we ask: Will a public awakening come sooner or later? Be greater or narrower in outlook? Have limited or universal effect? Be smooth or rough in transition?

If civilization is becoming an increasingly conscious, collective and unified endeavor—a civilization of great and determined outcomes—then let us make this civilization the first *universal* in scope, the first *democratic* in intimate social organization, and the first that is completely *life-fulfilling*.

We can put cheap, raw materialism behind us, use it as widely as is appropriate, but use it mostly as a springboard to a social greatness not yet seen on earth. That greatness cannot avoid being urban in scope, inspiration, and organization.

We human beings are our own creation. No other creature can remotely define its own form and destiny. We make ourselves through the society we build. Today we are making a completely new society. And with it, we are making a new social nature for ourselves. Can we do so with aplomb, magnanimous spirit, and high imagination? Can we broaden our vision to match the widest scope of society's fantastic abilities and continuing creations?

More Perfect Union

> *Human potentialities, whether physical or mental, are expressed only to the extent that circumstances are favorable to their manifestation.*
>
> Rene Dubos

If the radical form of progress we now experience has exploded the material things and processes of life, one day we may realize that this massive industrial outpouring might have been but the prelude for an unparalleled awakening. Ultimately, the larger human values of our transformation may be the vast opportunities and freedoms that industry will offer us—to pursue the greatest human possibilities in the many yet unexplored reaches of our social being.

That conversion of the historically new technical instruments of society into a more human content of life might be likened to the second-stage rocket that puts society into its first grand orbit of humanism. Upon the largely industrial, first-stage rocket of a million *technical inventions* that established the capacities for a new inspired development of society, a new second stage of modernization can then ignite a million *human inspirations*.

There is now, however, only a limited awareness, less expectation, and no clear momentum for this kind of change, even as there are now many disparate activities that may in time come together and form such a momentum. Today we can note, for example, a new vigor of museums; the many nature centers emphasizing education and ecology; the diverse air, water, and winter sports; the co-housing with varied family living arrangements. Nevertheless, most of the necessary creative personal, interpersonal, and organizational foundations for a new freedom remain sadly bound up in the race for raw production and promoted consumption. Although a new human enlightenment may be inevitable in time, certainly the achievements can be more assured and less faltering if the potential is widely understood, fully debated, and systematically organized.

Today we see many social movements arising in society. For the most part, the more visible public demonstrations reflect the younger generation's reactions to outdated customs and contradictions they see in gross material consumptiveness, racial inequities, civil rights, and war. But these movements are largely reactions against conditions in the larger society, not movements based on perceptions of the social potential. While the various civil rights movements cannot be belittled, their aim is a fulfillment or enlargement of existing rights, not the creation of new ones.

We underscore the obvious to say that people still perceive progress in technical and material forms. Private enterprise is understood as the vehicle to achieve most progress in society. Similarly, even *social* progress is perceived as emanating from enterprises that entertain, organize tours, or provide the finer things of life, like commercial art galleries and fine restaurants. Affluence, most people believe, is the single best, if not the only, avenue to good living.

However, as we note, the deep constrictiveness, grave debilitations, and outright contradictions based on the gross acceleration of consumption have made the good life illusory, empty, and self-defeating. Most of all, this model simply lacks vision, having locked into our present material aspirations from the age of steam. Our highly organized approach to raw production and consumption and our passive approach to the social dynamics of human advancement point to what may be the greatest blindness of our time. There is no inherent reason for such a lapse, only our stunted, economically dominated psyche. With new vision of life in cities, a new course of progress can be set, guided by a philosophy for strategic navigation and a tactical *social* use of the experimental method.

The brisk winds of a civilized awakening will appear partly as theory, partly as prediction and expectation, partly as public policy and planning, and certainly as social aspiration. Historically, material and scientific progress occurred when these factors interacted and gathered momentum. Social progress can do so as well.

The human transformation even now may be confidently predicted. To date, however, the pursuit of *social* imagination, philosophy, and leadership has been severely discouraged by the ideological and political stresses of the twentieth century and brushed aside by the onslaught of technical, corporate, and economic forces. Consequently, a direct and positive concern for the human results of change is fragmentary rather than whole, specialized rather than comprehensive, remedial rather than goal seeking, and reactive rather than creative. Such a process takes little account of social dynamic and loses individual focus in the anonymous world.

If in reality society has created a magnificent foundation for a profound and specifically social progress—a true renaissance plus technical power—what are some of the major underlying conditions we might expect or strive for?

1. A great plurality of human interests is already coming into being, exemplified by the rich, wide-ranging programs of Elderhostel and Earthwatch and by many other initiatives. Yet these are occasional enriching experiences and rarely reach home or regularity, where greater achievements and satisfactions may be expected.

2. A spirited human élan requires reestablishing a cohesive and dynamic locality for people within the larger urban framework, an integrated focal point for activities and services that encourages varied, regular, casual, face-to-face associations so essential to invigorated and personally fruitful relationships.

3. A high level of urban, social, and individual efficiencies is achievable and necessary to assure ease and convenience in daily affairs. Such efficiencies can simultaneously support urban spaciousness, varied forms of conservation, and many varieties of association and cooperation.

4. Cooperative behavior can be stressed over the competitive mode now dominating the public arena. Competition is demanding, abrasive, stressful. It requires losers and is best limited to activities that people may choose such as sports or arts contests. Possibilities for formal and informal cooperation can be multiplied by shaping facilities, organizations, and activities for that purpose, just as the organization of business has emphasized competition.

5. Personal interests and social ethics can be more closely united when the individual can find the security that permits easy association and involvement, build enduring and mutually endearing associations, and find honor in the eyes of friends.

6. Good works and creativity require leisure with a breadth of vision and varied stimulation. Today these conditions are subordinated to the race to produce and consume, the anonymous fracturing of the city, and the overwhelming emphasis on isolating individuation outside of a social context in which individuality can attain meaningful strength.

7. Rich, varied, participative organizations can become a strong incubating medium for personal growth when they are organized at the human scale, supported within a cohesive local setting, and integrated as a part of education, recreation, tradition, and public reward.

8. Traditions developed close to the person are important sources to build an inner pride, a comforting sense of identity, and a social loyalty.

9. Democracy can develop vigorous *social* opportunities and rights, comparable to political rights.

A century ago, these concepts would have appeared pointless when virtually all interaction was already at the human scale, when both work and consumption intimately involved the whole family, when the personal wealth to support personal expression was severely limited, and when long, arduous work was demanded. But today, economic wealth carries with it an increasingly burdensome overhead on the psyche that confines human growth and simultaneously blinds us to the promise of this most auspicious time in history.

Perhaps one day, society will experience a truly *popular* renaissance, not merely a civilized engagement among a very small, privileged elite. Most people can then be critically and creatively immersed in the greatest springtime in human history. However, this kind of human vitality can only occur when society gives the same dedicated attention to social progress that it now gives to economic expansion—diverse attention to finance, law, organization, personnel and training, product development, and marketing. When we do this comparably and directly for people, we may stimulate an incomparable renaissance.

Now, if "we the people, in order to form a more perfect union," is to be equal to the potential of our time, a new mandate is vital that will usher in a completely new order of social democracy.

Such an imperative is constitutional to its core. "We the people" speaks to the locus of sovereignty. Achieving "a more perfect union" highlights the organizational instruments necessary for action. Clear purpose is indicated in a later phrase, "the pursuit of happiness." Whereas the essence of the Constitution is of, by, and for the people in all of its parts and in the sum total of its provisions, contemporary social progress is severely stunted by the chaotic and confined pursuit of wealth, not happiness.

The American Constitution established an ethical foundation for the conduct of society. Now a new ethic is required to make that greatest transformation of all history into modern terms of the equality, opportunity, and justice that were set forth in the Constitution of 1787. The new ethic will necessarily reexamine the individual person and her or his relationship to society to achieve a freer, fuller rendering of progress in personal terms.

A new *social* bill of rights is therefore the logical, just, and historic imperative of our day. It may become an amendment to our modernity that will command public affairs in the way that the original Bill of Rights serves the nation. However, we dare not hold a shallow, politically powered, special-interest debate that responds mainly to today's dominant forces—powers that are the curse of modern democracy. Our challenge is to generate a fundamental debate that will create a beacon for change well through the twenty-first century. That challenge to our intellect may be as great as the challenge to our politics.

As it is, the social questions of life have been made secondary and even trivial to the economic. That is unfortunate, perhaps tragic, for we all know how our social being is confined, (or feel every day as a numbness of thought and emotion), even as it is close to the heart of our existence and to our sense of humanity. Yet we suppress such thoughts, set them aside as matters of personal bias, or dismiss them entirely in favor of the "greater" issues of politics, management, or investment.

Today there is little substantial or relevant debate about the higher human possibilities created by the new forces of the twentieth century, except in continuing fantasies of new technologies themselves. Furthermore, there is hardly a foundation for such a debate to occur. Both the content and the parameters of thought, as well as an underlying philosophy, remain essentially unformed. There are no organizations to start a debate, as there are in ecology.

Yet even without debate on the larger human questions of our time, solid steps can be undertaken by individuals and awakened organizations. Declarations may be sounded across the land announcing varieties of the new ethic, especially declarations that reassert the sovereignty and supremacy of the person and the social imperatives underlying all future human progress. Declarations need to be founded upon the recognition that human existence and all forms of higher human experience are inherently shared, interpersonal, and social. The core of human motivation, the heart of creativity, and even the devotion to one's own life, is social—deeply, profusely, inspiringly. The best side of our social nature is established by finding higher levels of sharing among us all.

Our social nature is sovereign and paramount to the course and content of future change. No longer can we tolerate the social subordination of ourselves to the economic means of life, as recent generations have done. No longer will the word *social* be primarily a preface to the word *problems*. Our social nature is basic to all human goals.

We have succeeded to an astonishing degree in developing a logic of science and technology, of administration and economics. These are fine instruments for society, excellent models of a new course for social development to follow, an advanced human and social logic to pursue the higher human purposes. Then we will be able to align the technical and managerial forces completely in the service of those higher purposes.

New Directions for Change

> The significant problems we face cannot be solved
> at the same level of thinking we were at when we
> created them.

Albert Einstein

All too easily one may become disheartened, even cynical, about the mismanaged material force of progress—and about the prospects of reordering the course of change—propelled as it is today by unchallenged economic power. As an ideology, the free market system stands triumphant and unquestioned. Corporate special interests today are able to monopolize the direction, content, and force of change throughout society—almost as a divine right of free enterprise. To be sure, the diverse creativity of small business and the enormous technical strength of great corporations are socially desirable overall. But they are also politically potent in promoting an unending, even reckless, expansion of production and consumption, whatever the social consequences.

The imperative of our time is not merely to set forth new priorities. We must reconstruct the very paradigm of thought, which has dominated society for the last two centuries—and to do so consciously, systematically, and with singular resolution. Such a demand will entail a thoroughgoing social self-analysis, virtually a collective Freudian psychoanalysis, resulting in a new formulation of the content of a newer and far freer experience of life. Since we are talking about the underlying values in which thought is formulated and behavior is organized, the challenge is unprecedented. We have to bring a totally new character of thought into human affairs, both to solve critical issues and to capture the new human potential.

Yet society today is dynamic and changing, and corrections can capture the leading elements of that change. Nevertheless, a more creative, socially valid direction of change cannot occur through the initiatives of the market and the responses of consumers. That part of the system has run its course as the leading force of change—and, indeed, as we see so many ways, is becoming socially disintegrative. The prevalent source of creative change must shift to people themselves, and this means that the source for new directions in life must center upon the democratic process, now more than ever the heart of humane progress.

How, then, are people to perceive and be inspired to redirect the leading edge of progress? If the indications of this book are even roughly valid, two sources of popular understanding are necessary to fundamentally redirect how human progress is understood, inspired, and undertaken:

1. The people must understand the destructive nature of cities as they have been built. Most critically, people must understand the role of the runaway economy in promoting the destructive form and functioning of cities. What will another two or three decades of a continually accelerating runaway economy do for our psyche, our struggle for dignity, or our human ecology? Somehow, that massive lesson must be learned deeply in our minds, our behavior, and to the roots of our soul.

2. The second and ultimately greater lesson is to spawn a legion of social ideals, emphasizing (a) the great possibilities of life in cities, (b) a broader, more social concept of democracy, and (c) a new, congenial, and permanently workable relationship with the environment.

In the end, both understandings become one process of a new perception of progress, no less than the flight of immigrants from abroad derived from undesirable living conditions on the one side and the inspiration of new freedom and new opportunities in the American land on the other. Crossing the ocean psychologically prepared them for new visions of life in the new world. Today we need a psychological break from the endless and increasingly debilitating struggle now in its watershed period of endlessly increasing simple production and consumption. Recognizing the historically unprecedented disruptions of misplaced output combined with a new vision of life's possibilities, society can inaugurate the essential transformation for a new course of human development.

Today the profound challenge lies in freeing people from the grip of economics, a grip that emerged with the development of industry and an intensive and dominating money system. That power has now shifted to a commercially commanded economy in which a massive profusion of advertising promotions in the still new electronic mass media is succeeding in reshaping the content of modern life to an unrelenting consumptive engorgement. This new power over society is permitted, surprisingly, by the rise of affluence and freely expendable wealth. Instead of freeing us, our affluence is used against us, binding us to a powerful discipline to produce more so we can consume more, thus denying a truly free social evolution of the human potential.

As long as the dialogue of change centers almost solely on growth economics, the human goals of change will be simply and utterly defeated. It is first necessary to establish a *social* legitimacy in local and national affairs. That is unfortunately necessary in a society not accustomed to giving social direction to the management of material wealth. Yet that is imperative in an affluent society, especially one revealing the disintegrative effects of the runaway economics. A new agenda must redirect the values and goals of our public debate. Shifting the debate from economic to social goals will mean a progress in all departments of life, possibly including some economic growth. Such a debate can—and must—

result in raising the social possibilities logically and powerfully toward a new era of progress. Here, then, is what I believe to be the intellectual and democratic challenge for the coming generation.

A profound public inspiration will happen, first, when we recognize the tragic debacle arising from industrial development and, second, when many examples of new social possibilities begin to demonstrate their infinitely rich, practical, environmental, personal, interpersonal, and cultural promise. Such an awakening, I believe, can inspire a new civilized spirit.

Organization and Locality

> *[The polis should house] the largest number which suffices for the purpose of life and can be taken in at a single view.*
>
> Aristotle

We cannot dismiss the possibility that the cherished American traditions of privacy and individuality, as worthy as they are, have become devices setting up the public masses for exploitation. They are, an any case, a spent myth now subordinated to rigorous business development. So perfectly do privacy and individuality fit the isolation of the suburban dwelling, the insulation of the family automobile, and the ideology of consumption, they virtually define the contemporary commercial goals now dominating the American Dream. Yet, accompanying these material profusions, are the mass appeals of professional sports and media entertainment—and the power of billions of dollars that underwrite the sponsoring-product promotions. Can it be that the tradition of rugged individuality has been reduced to the selection of products promoted by advertising?

Today, while bureaucracy overwhelmingly organizes society, our social values and personality are tightly subordinated to those organizations, so completely that we can say that the higher the social value the stronger is the organizational force surrounding it. A century ago, industrial workers were exploited until the union movement established counter organizational power to face up to the industrial giants of the time. Now we take it for granted that social strength is organizational strength, a strength able to effectively influence if not dominate a sector of human behavior.

And thus today, we have become a society of endless organizations serving endless special interests that may or may not coincide with the deeper human purposes. Many thousands of organizations serve industrial, trade, labor, and professional interests. Thousands more serve religious, cultural, recreational, sport, hobby, and conservation causes. Despite the many values served by all of these institutions, none serves the whole human being in the broadest range of personal and social needs.

So another kind of organization is conspicuously absent, or is but fragmentally established. That kind of organization is similar to society's oldest, found in hunter-gatherer and pre-industrial agriculture: clans and village communities. These were small, local, and comprehensive and served the well being of the whole person. The welfare of the individual was substantially shared by the whole group. Today neighborhood houses and community associations exist, but they are scarce, incomplete, and lame in broadly serving the whole individual and whole urban populations.

Instead, throughout metropolitan areas, we see a disarray of interpersonal and neighborhood values roughly reflecting the disarray of neighborhoods. There is an isolation between neighbors living contiguously, a fundamental inability to cooperate, a nearly total absence of vision for local urban living, and a deep-seated fear of becoming too close to other people. This tragic dissociation exists close to the cradle of personality and the associative core of human behavior. And this invisible wall separating people stands with us at a time when we can and should be elevating our social values.

As it is, families must leave their residential areas and go to dozens of urban locations to obtain the benefits of life. A person must overcome a geographic hardship to obtain needed products and services or participate in recreational and cultural activities. We see a great deal of organization in creating the good things of life but very little organization in making them available, close, and convenient to people in their home settings. People are at the end of the line, so to speak, unable to shape the things and behaviors of life for optimal living. They are forcibly isolated from these benefits of life, especially by zoning to systematically segregate and isolate urban functions, as well as by the happenstances of urban development. Both public and private forces literally disintegrate urban behavior of people.

Clearly, society gives little integrity to persons in their living habitats. With all of the varieties of modern organization, we do not reaffirm that life is and must necessarily be finite, local, and integral to the citizen. Crime, for example, is not a set of national statistics but rather concrete local events with local causes. Yet if statistics have become true reality of society, it is probably because the localities of life have become barren, discordant and threatening, and for so many, a life of either creeping or dramatic tragedy—the same general conditions that is or promotes crime.

If we ignore the social integrity of locality, we ignore the integrity of the person. Then we look at statistics and lose the realities of both persons and localities.

Therefore, the statistics of crime can only grow. Yet building the integrity of localities and persons is not an anti-crime strategy. It is much, much more. Foremost,
it is a human, local, interpersonal integrity that will serve the positive human
aspirations.

Our age is indelibly marked by universalized functions in the big society that
overwhelms the unique person and place, which characterized human behavior in
the past. That is, society makes the engagement of each of us universal to the
whole society as much as possible. By telephone or television, jet or automobile,
one has ready access to infinite personal interests or public events, be they business, news, sports, or entertainment. The revolutionary pace of electronic communication is broadening that scope. Thus increasingly, we pursue our many
special interests across town or across the country.

The trend is decidedly national. One can order a sweater from any part of the
country by phone from home with greater ease than going to the nearest regional
shopping mall. We can cross the country on the interstate highway system without stopping at a traffic signal. Even the home teams of the major leagues today
command less local loyalty in our highly mobile society than they once did.
Many persons are now as likely to root for a team that is a thousand miles away.
Television increasingly departs from locality, even in news and weather. Our
transactions are closed with international credit cards. Plainly, public interactions
are shaped for great scale and long distance, that is, to maximize the universality
of our involvements throughout society.

This nationalization of human behavior stresses functional, non-personal
human relationships. The individual must constantly negotiate the maze of complex bureaucracies amidst the anonymous labyrinth of services one seeks.
Therefore, the roots of local, interpersonal, and continuing relationships are a
greater burden, become less frequent, and often merely dissipate. One may
become better acquainted with a person across country than with neighbors only
yards away. This deterioration of the place in life is symbolized in the fact that
walking and the walker, face-to-face interaction, and association for its own sake
have no public priority and carry little of the integrity of the national media.

Our question today is this: How are we to conceive of a kind of living in
which both the practical benefits and the most comforting social arrangements
are organized of, by, and for the residents in a close, socially ample setting near
one's home? History offers us the pastoral community, which, for all of its inade-
quacies, demonstrated a completeness and human scale of behavior founded
upon the local functions that maintained life. Today we apply the term community to a wide range of common interests in a comforting empathy with that history and usually with some sense of communion with others. This, it seems,

reflects a wistful longing for a level of human association of which we are now seriously deprived.

To give the person integrity and power of organization in the modern context, some very specific conditions come to the fore. The first is physically uniting a range of people with a range of commercial, public, and cultural services. Second, this may be best when it achieves a limited but personally manageable diversity for individuals, a scope of activities that is neither too small and confining nor too large and overwhelming. The all-important feature is human scale: physical, social, and institutional. From this point we can begin to define a modern, achievable community with an immense social and cultural potential. With vision and will plus a growing range of experience, such communities should be able to far surpass the human benefits of traditional community.

Community is an organization that can serve the whole individual, that is, serve the person as person and not merely provide services. Its object is to give the person the same vital organizational support that is enjoyed by other sectors of society. The community can restore some of the individualism now idealized in myth, especially to give personality its own optimum freedom, strength, and richness.

Somehow our modernism has created gaping holes in our democratic system. One is the space closest to the individual, the place where ordinary daily events might build more healthy, trusting, and sharing relationships. That can be the place with more immediate opportunities, and the place with a new, particularly modern freedom afforded by our great social wealth. But now these values are effectively and ironically submerged.

Democracy is increasingly meaningless when it foregoes pursuing a progress of its own, independent of economics, and parallel with other arenas in society, capturing and incorporating their vitality, and building the great social possibilities inherent in our time. For example, the historic town meeting has no counterpart in urban America, not because it is outdated but because the structural basis for its evolution was never created in cities. Municipal government was thought sufficient, but it, too, lost touch with individuals and group life. A true community had no basis to evolve over the decades within the new circumstances of cities. Only a democratic organization close to the person and at the human scale can establish a sufficiency of social integrity to balance the massive, anonymous, dispersed, and powerfully organized mass society.

Essential Community Integrity

Like a person's body, community requires a unity to function coherently and freely for its members. That unity demonstrates an integrity common to all organizations. The foundation of an organism's unity and integrity is the cell, the smallest element capable of independent functioning, which reveals its integrity. Similarly, products, the factories that produce them, the corporation, and government bodies demonstrate a unity and integrity of form and function.

However, in today's cities, there is an absence of organized social integrity, locally at the behavioral level

Toward Creative Cities

> We cannot undo city life, we cannot prevent it, but perhaps we can humanize it.... I should like to see the city of the future broken up into a large number of small self-contained sub-units.... These ought to be small enough...to be personally acquainted.

Arnold Toynbee

In the previous chapters, we reviewed the historically critical urban predicament of society, emphasizing structural and comprehensive problems. While the current form of the city decays in its essential ability to serve the person, cities overwhelmingly respond to their severe problems largely as questions outside the city's form, keeping to narrow, specialized actions like crime suppression, remedial education, or air and water pollution. These specialized, issue-by-issue programs fail to resolve or even acknowledge the multi-issue, common-source, physical-unity problems imbedded in today's cities. A measure of success of a specialized program, for example, may alleviate traffic congestion or reduce crime locally or temporarily. But this covers over and may even aggravate the deeper, long-term destructive turmoil built into urban form by traffic and crime.

The vast urban problem demands profound rethinking. Simultaneously, the vast, generally unimagined potential of cities calls for equally profound thought. Both the negative and positive sides of the urban question are bound up together and demand a new unified theoretical foundation for cities.

We will therefore consider in the following chapters a broad range of concepts pursuing those goals. As far as possible, I try to define basic or ultimate possibilities, not partial, step-by-step actions. There are two reasons for this: First, step-by-step actions are not likely to demonstrate the greater urban potential, and this is critical if the greatest possibilities are to be visualized and if society is to consciously build social ideals into the formative structures of cities. Second, many of the concepts I set forth are not amenable to step-by-step evolution unless an ultimate urban form is understood and shaped into the initial planning.

A new and fully comprehensive approach to urban development itself might be considered the first principle encompassing social, economic, ecological, and cultural objectives, as well as the physical. A good city cannot be created when particular objectives are viewed independently of the whole, such as those relating to transportation, commerce, or industry. The highest level of integration is as necessary to the city as it is to the human body.

All societies organize best what they prize most. This, too, might be a principle to be applied to cities. In basic ways, and especially in human terms, the modern city is now unorganized, even disorganized. Today, with no clear ideal of the city, there is no aspiration for higher urban achievements. Consequently, there is no applicable or effective theory of cities. Most of all, private initiative and public regulation together offer no useful, organized way to build a unified city. Nor is there now any organized basis for cities to serve the whole person. Municipalities, like businesses, are organized by services rendered (not for the people served), most services being an independent entity independently administered. Cities will fail deeply until the person is made organizationally central to urban development.

In our day of mass machinery and pervasive corporate organization, sound urban theory must master both technology and bureaucracy as they serve people and operate in human environments. If one day we prize cities as we prize sophisticated technology, those cities will effectively shape both production and product precisely and completely in people's favor. In short, we need to give the complexity of human needs a completely new order of public concern and make them supreme in public affairs.

Without question, our age has developed the machinery for a new civilization. We must now turn to ourselves to achieve a full, human flourishing civilization. The technology available to us today offers so much and so many kinds of opportunity. But we see so little of the greater potential. The more significant realms are what is made humanly possible: the new social range of freedom, building greater strength of personality, finding new realms and uses of art, creating an unlimited range of interpersonal dialogue and behavior, and striking new sources of inspiration—altogether, expanding the many arts of living. All previous civilizations may then be considered partial, elitist, specialized, autocratic, regional, and terminal.

A major transformation of thought, as I repeatedly suggest, is necessary because cities are not a product, either consumptive or durable, although we now build them as if they were. Cities cannot, therefore, be conceived strictly in terms of economics, even as they obviously do require economic resources. But there is no current organizational way to build them except as a simple, for-sale product. So the city is organized for its economic functions, especially the housing market. But regardless of how we may view human integrity, social adequacy cannot arise from salable or alienable arrangements.

By custom we think in terms of organizing particular functions for their separate operation and efficiency. Supermarkets, shopping centers, schools, hospitals, and airports are organized in great detail for smooth internal operation to provide a special service to the public. But as individuals, we are overwhelmed by the dizzying variety of these services and by the burden and complexity of obtaining them. That is because the paramount integrity has been built into the very essence of these separate institutions and their separate services, which makes them overwhelming and burdensome for people.

To truly serve persons, we need to combine as many organizations and their services as we can that are directly relevant to the person. Then we must integrate them in one place for personal efficiency so that one may use any or all of them at one time. Then they need to include not only commercial services, but also a comparable range of governmental, social, and cultural activities. Finally, they need to be as close to the public's residences as possible. Old town centers came moderately close to this ideal. But shopping malls clearly do not, for they are not public, not publicly controllable, and most of all, demand the automobile for access.

Therefore, society must call for a mammoth seismic shift of its thought to move from the centrality of specialized services to the centrality and paramount position of people. Service efficiencies need not suffer and indeed, should improve. Functions will then become integrated dynamically around people, serving their needs, their sensibilities and scale, their personal efficiencies.

The profound urban discontents and the great urban promise both argue for a vigorous new way of looking at ourselves within the environments we build— which organize our behavior, thought, and personality. Since cities organize the whole of civilization down to every person, society cannot risk leaving its urban creativity to the massive technopower of the runaway economy. Nor can society continue to suffer the structured destructiveness inherent in today's urban form. Cities require the central place in public action because they establish the constitution of our physical and social behavior.

What follows is a broad span of integrative ideas formulated as a general theory of urban form. That framework will seek a fresh human logic of life in cities that is socially expansive, culturally creative, personally secure, publicly efficient, newly free, very urbane, close to and comfortable with nature, ecologically sound, and demonstrating an economy of least means throughout.

CHAPTER VIII

PUBLIC-INTEREST IDEOLOGY

Three Organizing Systems

Historically the two basic systems organizing society—democracy and capitalism—emerged with the country from its birth, one by a grand stroke of creation, the other by vigorous evolution. Both have thoroughgoing ideologies and loyalties that merge intimately with patriotism. They were involved in the two greatest threats to society in the last century, the Great Depression, an economic crisis, and World War II, the greatest world crisis of modern history. Both systems were also engaged deeply in the Cold War, the four-decade standoff between the democratic and capitalistic ideologies and the totalitarian communist system. Both Democracy and capitalism today stand exultant from successfully weathering and triumph as we proceed into the new century, having evolved into what can be regarded as the most dynamic society—free and prosperous—in human history.

However, these mega facts of the last century also reflected a course that, while on the one side generated an unprecedented wealth for the majority of people, on the other left the people in very unfamiliar and debilitating psycho-social conditions, undermining traditional and close interpersonal behaviors, and setting society on a course of ecological self-destruction. All too clearly, capitalism has shaped society in its own interest more basically than for the people they were intended to serve. We are a society of huge and enormously successful corporations that guide research and capital to serve profits more powerfully than the public benefit, most poignantly in health care.

Merely to take note of the many revolutionary developments is to perceive the fierce social storms that have enshrouded the individual person since these forces set out after 1776 and 1787. We regularly read about the Industrial Revolution and the generation of abundance. Far less do we consider the socially disruptive impacts of those same forces upon the psychosocial roots of organized life. When we do consider the generation of social crises, they are normally cast as isolated social problems to be met pragmatically as one issue requiring one programmatic solution, whether family breakdown, criminal drugs, or a high rate of homicide. They are treated as issues unrelated to the immense forces of our history, that is, hardly at all connected to the massive social transformational side effects resulting from the Industrial Revolution.

These immense transformations through which society has passed are as basic to of the Industrial Revolution as is the increase in productive wealth. Here I set forth some examples of basic and corollary developments since 1776 now impacting the human side of society:

I. The rise of large basic industry late in the nineteenth century, *then mass production* in the first half of the twentieth century, *and finally automated production* late in the twentieth century.

II. The growth of cities, from hardly more than the very small cities of Philadelphia, New York, and Boston in 1776 to a society totally dominated by the mass urbanity today.

III. The emergence of corporations from seedling entities two hundred years ago to become the characteristic and dominant form of economic organization we know today.

IV. The appearance of mass marketing as an outcome of mass urbanism, mass production, mass media, and universal advertising under the direction of the corporations.

V. The explosion of varied mass media from a few dozen small presses to an immense system of book and periodical production, followed by radio, film, and television, and now taken up by the Internet.

VI. The penetration of advertising into most media on a scale that is massive, sophisticated, penetrating, insistent, and unrelenting as a prevalent social tool aimed at influencing if not dominating social behavior.

VII. The development of jet travel and the automobile, the first dominating transcontinental travel and the second dominating the form of cities.

VIII. The creation of instant, versatile, personal communication by telephone, including the cellular telephone, fax and, again, the Internet.

IX. The creation of massive, universal bureaucracy, substituting a mass functionalism for much of the most common historic human relationships outside the family.

X. The creation of a system of R&D to regularize inventiveness by, discovering, reordering, and composing knowledge to develop new products and processes for society.

These ten arenas constitute the greatest revolution of human history. None existed significantly at the founding of the republic; indeed, they were still primitive in 1900. They underwrote the growing standard of living yet, more importantly, became the essential ingredients organizing living. *We didn't just experience revolution. We are its results.* Society became organized by its revolution. While filling the person's material needs, these forces also reshaped the foundations of thought, association, and action. But most significantly, the human outcome was incidental, derivative, and socially demeaning. Without intent, these forces redefined our humanity, creating bureaucratic competence throughout society but undermining close-knit social behavior. The material wealth, the new behaviors, and the new forces behind them constitute what we are today. People and society as a whole became unintended outcomes in which the numerous crises resulting were treated with band-aids.

At the same time, the historic qualities of good living dissipated without defenses, including much of the casual, informal, familiar, and indeed socially intimate behaviors. Thus, while the two systems shaping society were enormously successful in building the modern society, the economy creates a heightened anxiety of economic instability, while also creating a huge psychic crevasse in ordinary social affairs, leaving persons in social isolation. What is absent is a focus upon specific human needs, not consumption as such, a public value system to match and give new direction to the material progress we enjoy. We lack the social goals establishing a person-centered form of human organization, new and more unified approaches to urban affairs, and, perhaps most of all, a new vision of the person and of human well-being.

These ten forces represent technical changes in society that, while increasing the standard of living, have not otherwise benefited people in their most human qualities. A vibrant cosmopolitan life might be an exception, except that it involves only an extremely small portion of the population and is largely confined to the classic terms of European culture. So today society lacks a foundation to promote a universal social fulfillment, a system of human benefits through human initiative and social development.

There is in my view, therefore, an overwhelming case for the creation of a third major organizing force in society, one that serves the person *as person,* to build a social foundation for the good society, not merely a material amplitude of expanded private expenditures. People lack a clear means for many forms of purely social behaviors to flourish; economic benefits, for all of their values, are not enough and, indeed, have become errant and interfere with associational behaviors. People require a new unity in social affairs that also affords wider personal options in life. It is not enough for the person to seek psychotherapy and to gird the loins to meet the next series of the alienating demands associated with producer-consumer behavior. *A third organizing force is needed to set forth new and greater living values in the organization of life.* Democracy as now practiced is insufficient. Once affluence is reached, the power of economics to twist human behavior becomes socially dangerous. While the person has benefited from narrow economic performance and remains theoretically supreme in democratic terms, the challenge for a third realm of social organization is to develop a new orientation that serves the person in a cohesive manner with new, clear, positive, social opportunities.

New Organizing Principle

The huge crevasse between the great material abundance and the vacuity of human social worth in the organization of society reflects a deeply inadequate philosophic approach to modern development that could have led the way for a different course for the great changes that occurred through the last century. At the same time there has been a huge development of corporate and governmental organization that has become seriously self-serving while filling specific human needs but left at bay the larger human purposes of life in society.

While society serves its members for the bulk of the functional details of well-being, it has not pursued the human integrity and inherent social dynamism of people, a level of sophisticated organization that is so highly developed in capitalism and democratic government. Society has learned how to secure the political sovereignty for the citizen and material abundance for the consumer, but left out the essential ingredient of our most fundamental social nature. *One might say that we lack a social dimension to our democratic practice and a social redirection for our material abundance.*

These highly successful, yet hugely biased organizations of life are readily seen in both the achievements and failures of society in the twentieth century: the immensely wasteful wealth that overbuilds and destroys cities; the institutional power reflected in the purpose, organization, and methods of education; the degraded processing of workers and consumers to promote production and marketing; the economic determinism reflected in corporate power in governmental affairs; the management of health care for its proprietary interests; the regularizing of crime as a growth economy and permanent public affair; and the deep cynical and paranoid character that has become embedded in human relations. These successes and failures merely reflect the deeper absence of a humanizing force at the core of society. At worst they represent a development by destruction, at best a wastefulness arising ironically from our work ethic.

The most unfortunate nature of this most modern debacle is that the massive interlocking forces and their multiple consequences are not recognized in their essential wholeness. In our narrow specialized perceptions and pragmatic manner of response, we recognize the parts and pieces of each particular issue and apply direct and narrowly focused solutions to resolve them. We fail, however, to recognize the holistic character of people and society, their issues that are part of the total historic change or the completely interwoven nature of nearly every issue. More fundamentally, we do not pursue social ideals and organize our history for all that a good society could offer. We fail to perceive the deep human content of progress that awaits the recognition of our most dynamic human potential. Both capitalism and democracy could evolve because they both had solid foundations upon which to evolve: From Adam Smith through Alfred Marshall to John Maynard Keynes, economists developed the doctrine upon which capitalism advanced. John Locke and other Enlightenment thinkers created the ideological foundations that were set into the U.S. Constitution. Yet society now stands naked against the broad potential for an expansive humanism, which is without theory, method, overall organization, or even a name to establish a common identity.

We are compelled to ask, then, what is the core principle that might serve the needs and greater human possibilities particular to our time and aspirations? My belief is that there is a rational scope for an organizing force that might be established as the *public interest*. This force would build systems to serve our fundamental social nature and pursue the greater human potential. The public interest as an open-ended concept focused upon human purpose, can achieve a clear ideology and create an organizational framework for common public action. Included are those things that directly impinge upon each person, especially the freedom found in readily available settings and activities that the person can effectively choose his or her relations far more completely with other persons. This is indeed an astonishing shortcoming of society that supposedly protects people as sovereign in a democracy. But in a society intimately shaped by established organizations, the public interest can create socially receptive and responsive organizations for a greater humanity to flourish. This is the inspired challenge now confronting society.

How might we, therefore, define this new principle to build the public interest? It includes, most of all, I believe, two spheres of human behavior, first the local, general-purpose consciously created urban communities that might become basic organizing elements of cities. Second, the public interest would include the many specialized voluntary organizations already serving many social interests of society. In both cases the sole objective is to promote and guide social development as determined by the people. The public interest includes all public and private services that directly serve the people, such as consumer economics, health services, human environments; support personal self-development and assist learning throughout life; develop personal opportunities to expand significant human experience, thus effectively expanding inner and outer freedoms.

The primary interest is always the well-being of people in vital human association as each person determines.

The most effective development of a new range of freedom, I emphasize, stresses varied physical, social, psychological, economic, and cultural opportunities available to serve the person. *Community would offer new broad ranging opportunities that constitute a new scope of freedom.* And community has numerous possibilities that will simultaneously serve specifically human economic and ecological purposes, being built around the development of a universal efficiency in organizing human affairs. And in keeping with the individual person to identify with socially important organizations of society, each community presents possibilities that are social, positive, and local to build economic security, social opportunities, psychological identities, and personal élan.

This means that community shall underwrite a clear, stout self-development of the person. A self-development of each and all members of community, as well as each community itself, creates a process and an air of human growth beneficial to every person. It sets the tone and organizes the best means to elevate the individual person to the highest purpose of society.

In so doing, the public interest offers society the basis for a public philosophy to be formulated and expanded to the fullest human conceptions of life, offering a concrete basis for personal and social goals to be defined, adopted, and set into motion. From varied philosophies, a public-interest ideology can evolve alongside those of democracy and capitalism.

If cities constitute the overwhelming scope of modern living, they deserve a new philosophy establishing that purpose. But today, as I frequently stress, cities seem to be noticed only for their problems, not their potential. This must end. If one branch of the American political parties identifies closely with capitalism, the Republican party, and is associated closely with its ideology, it is reasonable to project that the other branch, the Democratic party, might pursue a full development of an urban and community ideology, while the ideology of democracy as a whole spans the entire range of the political spectrum.

The specialized service organizations are already very extensive, including such diverse activities as the Red Cross, Boy and Girl Scouts, YMCA and YWCA. It also encompasses many other public interests such as Common Cause and Public Citizen; conservation groups such as the Sierra Club, Audubon Society, Nature Conservancy, and National Wildlife Federation; civil rights organizations such as Amnesty International, Human Rights Watch and the ACLU. All churches, lodges, and membership organizations are a part, as are all direct-service medical and educational institutions. The sources of funding are highly varied, including public and private contributions, as well as earned income. Some private universities today are receiving significant public funds, such as Temple and Pittsburgh universities that are classified as "state related." Conversely, many public universities are now receiving larger proportions of their income from private sources.

What is basic is that all public interest organizations are their services to people or their membership of people.

What may be preventing a common perception and outlook for a public-interest structure of society to build upon is the very breadth of its organizations, services, and traditions that could constitute the third major organizing sector of society. The scarce current range of urban community organizations and the barren plight of the person in modern urban society indicate that even a common outlook in that field does not exist. Cities have no movement comparable to that of the conservation of nature, an ironic fact in the face of evidence that so many of the problems of natural ecology arise from the disastrous consumptive forms of cities that are not amenable to individual conservation practices. A union of conservation and urban interests could potentially establish a strong core for a greater public-interest movement to be formed. As it is, I have encountered only one relatively incidental, non-elaborated proposal for a balanced three-part system of organizing society. It suggests a "civil society…composed of communities and other collective associations," that together create a "stable tripod of governmental, civic, and private institutions." ("Globalizing Democracy," Benjamin Barber, *The American Prospect*, pp. 18-19, September 11, 2000) Despite the non-unified realm of the greater public interest, I would argue that the emergence of such a movement could be one of the greater social developments in the twenty-first century, especially if local communities can emerge to play a significant role.

What we can look forward to, I believe, is a greater, more articulate civil society constituted through a public-interest concept. That system can be developed as completely and forthrightly as democracy and capitalism. Its purpose, organization, and processes will, of course, require a thoroughgoing political theory, common ideology, social methodology, and much experience to set it into practice across the realm of public affairs.

That justification carries profound human possibilities. If we look upon nature, the prevailing struggle is the Darwinian survival of species; the rise of civilization shifted the struggle for human survival to the human groups that have grown ever larger and are now becoming worldwide. In our time, effort (excluding war) has shifted to amplifying good living, reflected in the rising standard of living during the race to produce and later to consume in abundance. Now the challenge is for all people to prosper in all categories of good living, which is clearly the higher prospect of civilization. That last stage now coming into view is different in purpose, different in organization and method, different in economics, and different in the evaluations we make of life itself. That stage logically dwells upon the individual person as never before.

Public-Interest Economics

If human progress as we know it occurred largely through immense achievements of technology and capitalism, we can look to this wealth-making arm as the chief foundation from which a higher humanity might evolve, especially through a public-interest system. While during the last century capitalism has bound society to the powerful economic necessity, displacing the necessity of nature, we now need to look to economics not only to provide the necessary material foundation but also to reduce and possibly eliminate the economic necessity. If the public interest is to focus upon the supremacy of person, it is necessary to relieve society from the domination of aggressive economics, paradoxically providing relief from the runaway economy.

Whereas today's economy is divided into private and governmental sectors, a logical reform would add a public-interest sector to convert a reasonably large share of economic wealth to create new dimensions of social and cultural value. That wealth would translate into a large expansion of opportunities for individuals and groups to pursue their varied interests. The existing economic capacity to underwrite such an inherent human growth is enormous, ironically best demonstrated by the demands of wars and the waste generating runaway economy itself.

While capitalism has grown to dominate society, this power has grown mostly without a social competence or democratic justification. Indeed, through social fragmentation of consumers and the universal pressures to produce and consume, the human ideals tend to be lost in monetary exchange, the rigid economic ideology and growing waste behind the once valid standard of living. In total effect, this commercialized ideal plus the growth of fragmented mass consumer society, has all but denied a social coherence great enough for society to visualize higher social ends. Commercialized consumption has been crowned the monarch of the social realm and the corporate economy jealously defends the commercial ideal against all challenges, especially those that can be branded as "big government" or "tax and spend" or "socialistic."

The issue is simple and nearly absolute. Whereas capitalism cannot be socially or legally responsible for any matter beyond marketing products and making profits, its grip upon politics and its ability to regularly defeat public bond measures, for example, means that capitalism can politically veto measures to liberalize the organization of life, and remains supreme to this day. Hence, while capitalism is by both law and competitive doctrine incapable of filling broad human ambitions beyond purchases, its political presence prevents the pursuit of higher human values that its own productive genius has made possible.

The powerful autonomous actions of corporations present society with a massive historic conflict of interest, a conflict between economics that corporations control and the broad uses of social wealth that they effectively deny. Despite the fact that the economy is the instrument of society, to be organized as society determines, the corporately controlled economy and the tight leash that it has upon society means that it can deny the free exercise of social affairs.

The greatest evidence of the conflict of interest is seen in the size and nature of the runaway economy and the forces that generate its spurious dynamics. Being a measure of that conflict, the runaway economy—above and beyond a high standard of living—prevents a greater social purpose from coming into being. Most simply, capitalism preempts an independent course of social development.

Thus today we live within a democracy of proud traditions where, contrary to the freedom of social development, a virtual vacuum exists in the realm of social development—the arena in which much future human progress will inevitably depend. *The public interest today is lost to the counterproductive wilderness of the runaway economy.* What is all to evidently necessary, therefore, is a broad and penetrating social dialogue that can build visions of what a public-interest system might do. That dialogue might create a social perspective of *a transition from a society organized by the productivity of capitalism to a society organized by society's creative uses of wealth.* The promotion of consumption, by itself and beyond any reasonable value, is an illegitimate corporate power and therefore stands in fundamental conflict with the higher interests of society.

One may rationally ask on constitutional grounds whether a system of economics is licensed to *govern social change*, and do so by autonomous corporations serving only their profit-making purpose? Can they *reform personality* in the image of compelled mass consumption, substituting their socially alienating market processes for a socially coherent, democratically determined direction of public affairs? If we continue to promote the endless growth of socially distorting production and consumption, we inevitably confine our visions of personal and social life to inane output and gluttonous engorgement that is ecologically unsustainable.

The production and consumption equation was legitimate until general affluence prevailed, at which point it turned wasteful, then destructive. So no longer is the economic equation of the market a worthy human model, especially where it governs social organization and personal behavior. It merely deflates the human spirit, in which a growing and pervasive cynicism is regularly demonstrated in public opinion polls. These are not incidental or momentary reactions, but are embedded in the logic of indefinite economic expansion by forced market consumption and disruptive materialism like outlandish automobile charges and intolerable congestion on billion-dollar freeways.

Affluence has now been achieved for roughly three fourths of the American population. The challenge now is totally new and very different from the struggles for material sufficiency in times past. Severe competition for scarce wealth is no longer singularly relevant as it once was; and demonstrative class distinctions are unworthy of the human energy they consume. Our urgent need is to establish cooperation and social cohesion as our primary social mode. That way of shaping the good life is a common public order in which the greater and possibly the greatest benefits of life are essentially public in nature and achieved through cooperation, a principle that in part supplants the presently overworked notions of competition, the seeming liberating but increasingly confining role of private monetary wealth. This was revealed by the magnificence in ancient Athens and Rome. Now with far greater possibilities, we can turn our creative energies to promote the many varieties of personal triumph through the many environments, facilities, and programs whose sources are public; they are open, trusting, cooperative, value building, and creative in a thousand fields. They do not require the immense private organizing energies now necessary to build public facilities through long and painful ad hoc efforts to get new public museums and galleries, or to find scarce and ill-fitting facilities for mediocre conversion to galleries, theaters, and workshops for the creative arts.

Our critical lesson is to understand how society is a cooperative social system created by people, how it is a reflection of our common organized behaviors and what we are as a people in our most promising stage of social development. Our challenge now is to perceive what we can be, as persons, as communities, and as a society. Our new dynamic potential and its essential requirements are what we must grasp. Then we can begin to think clearly about what we can be. Economics is but the engine to create the form we want our society to take. We, as a people, as a democracy, are the essence of a grand awakening.

The conscious development of a public-interest system of society signals to everyone that human organization of society is now becoming vastly more important than the economic imperatives that today so tightly govern society. The public interest exists where the profound and richly varied human values and aspirations await their release. With the economic wealth already with us, supreme human values can soar to levels never before possible. But such accomplishments are not possible solely through private money used privately. So, shall we permit the productive powers that created our social promise prevent its fruition?

With the public-interest potential we can project a separation of three sectors of society like we now separate political powers in the Constitution. Each will have a moderating or perhaps a leavening influence upon the other two.

Informal Democracy

Philosophers through the ages have been skeptical about democracy, if not outright fearful, as they noted the execution of Socrates in democratic Athens. They identified democracy with mindlessness, even as mobs acting with senseless passion. This fear was not entirely absent during the framing of the American Constitution, as when it provided that the presidents would be elected through electoral intermediaries and when senators were to be selected by the state legislatures. But over the last two centuries philosophic thought has come to trust popular democracy, especially as education and communication advanced and experience of democracy matures with creative stability.

However, in modern mass society new basic questions have arisen concerning the fidelity of legislator voting to reflect the will of the electorate, noting rather that massive election campaign contributions and powerful lobbying seriously overwhelm popular interests with great economic power. If the growth of the corporate economy has resulted in hothouse pressure politics, a very significant focus of those pressures centers on and distorts the political arena, compromising the integrity of the legislator's responsibilities to the citizens.

The irony is that the danger to the democratic process today is not the individual citizens acting with mob-like emotions; they arise far more from the corporate elite acting in their special interests. In our mass society, the key issue is how the immense powers of society are controlled. That is particularly noted on a number of questions, when one considers today how those powers are diversely expressed by building the permanent form of the city for momentary profits by real-estate developers, the advertising-driven power controlling the mass media, the content and character of education, and the pervasive doctrines of private enterprise.

In reflecting upon the need for citizen involvement in the necessarily large scale of political action, the issue reduces itself to the ways in which citizens may organize and play more effective personal roles. Political parties were never mentioned in the Constitution but quickly proved necessary in selecting and promoting candidates for office, and in establishing party platforms. While political parties are informal, they are vital to the electoral system.

However, political parties do not ordinarily reach down to the level of the individual voter. Something more is required to effectively involve the citizen. Citizens need clarity so their knowledge and understanding may grow, varied investigations may proceed, political opinion may be expressed, potential legislation may be formulated, and political support may be gathered. Voting for candidates and ballot measures is but the last formal step.

Even now, much is happening on a wide range of public interests. Citizens have been organized for many purposes: civil rights, education, conservation, health, women, youth, elderly, and minority issues. The best interest of society is to expand citizen involvement and increase its democratic value. Such informal organizations could become more valuable to society if they were involved in a public-interest system—and thus become a more vital part of the democratic political process.

Already many of these organizations are registered as lobbyists on the national level. As an act of the public interest, voter registration might be modified so that voters could check off their choice of citizen-based membership organizations that best represent their civic interest so that such organizations would receive, say, $25 dollars of public-interest funds per voter interest to augment dues and other income. Citizens would thus indeed vote with their dollars and the total dollars those organizations receive would also indicate their popular strength. Such support would encourage larger voter turnout. Of course, each organization would have to be qualified as a *citizen-based organization* and account for the funds it receives.

Then, too, communities themselves constitute a strong foundation for organized but still informal political participation at metropolitan, state, and federal levels. And when many communities unite for a political purpose, they offer another dimension to the informal processes that might strengthen citizen participation. They can then generate more positive social development in society. Communities in this way can energize the social grass roots of citizen behavior, uniting them to become taproots of effective democratic action.

The broad range of organizations that compose informal democracy may be as critical to the democratic ethos as the election process itself. They operate within the democratic spirit, would be already organized and experienced, and could be seen as the political arm of the citizens. The objective is to rehabilitate democratic politics and the place to do that is at the citizen level, first through the many specialized regional or national membership organizations and, second, through the communities that persons might be a part of.

Developing the Public Interest

The three systems together form a basic, logical, complete, and liberal framework and amplify the human content of the modern, dynamic, and open society. The division of functions is clear:

Political: Democracy by exercising the people's sovereign power.

Economic: Capitalism as the fundamental source of output.

Social: The public interest as the organized basis to increase human fulfillment in the large, advanced, powerful, and wealthy society.

For human fulfillment, the prospect is that personal self-development (best within community) represents the highest achievement by the person, and the person achieving the highest level of virtu represents the highest achievement of society. Community represents the chief social means to nurture self-development, which therefore sets forth the central essence of the public interest. While other institutions and courses of development are available, community potentially offers the most direct and substantive scope of the human interest.

In instituting the public interest, the current terms of "non-profit" and "non-governmental" (or NGO) need to be replaced in favor of the "public interest." Those terms need to be reduced to what they are in reality, not be labeled merely by their descriptive modifiers. Both terms tell what the organizations are not, and are therefore negative descriptions that merely indicate the indifferent status of their roles in society. A term that is fully equivalent to democracy and capitalism is essential to the important roles of the public interest, potentially a very important course of development in this new century. The public interest can truly pursue human fulfillment—while also creating ecological sustainability and maintaining natural diversity.

If society is the created instrument of the people and civilization is their high achievement, an important part of that progress can be, I believe, assumed by the public interest. Already, without formal recognition or conscious ideology, it is a large contributor to good living, if we count the universities, schools, museums, galleries, and hospitals. And, allowing that public interest organizations can lead in creating articulated goals for society and pursue them diligently, their potential can only grow.

A general appreciation of the public-interest system will grow when we begin to perceive its purposes, objectives, organization, methods, and its record of achievements. New perceptions might include (a) associating the word "social" with progress and no longer followed by terms like "problems" or "crises" that now so denigrates our human perception of public affairs; (b) separating public-interest budgets from classic government operations; (c) building budgets around positive social programs (minimizing the current stress upon single-issue, single-program responses to crises); (d) establishing applied research centers that focus on the many sides of social development, that is, for projects to be set into state and national priorities; (e) progressively defining higher human health, vitality, and experience to be incorporated into the dialogue of social development; and (f) establishing national and local statistics to cover all measures of social conditions and progress.

Of critical importance to the many branches of the public interest is to build systems of public interest initiative for approved social goals, initiatives now basic for corporations. Heretofore, independent initiative is taken mostly by autonomous corporations that also hold a vice grip over public programs, especially when taxes and revenue bonds are involved. The precarious nature of even the most basic educational funding is the telling force of that grip on today's public affairs.

The historic lack of public initiative in social realms reflects what may be called the corporate ideological monopoly over the direction of change in society. The rationale behind corporate control has been the growing *private standard of living*. But now, if one accepts the line of reasoning herein, the initiative for human progress now logically shifts, at least in part, to the public-interest arena—or, one may say, to a *public standard of living*. But by the business domination over society throughout the last century, emphasis was always upon private market-based expenditures, and positive public-interest initiatives have been nearly closed out. Now, however, a new balance of progress and initiative is foreseeable. So the old excesses may be recognized for what they have become, an economic control of social affairs. Consequently a realignment of the ideological foundations of society is becoming both more urgent and more promising.

And thus, with an attitude shift that underscores the public interest, and possibly a reduction of the concentrated special-interest, money-in-politics, a broad shift to liberal public-interest initiatives is possible. New public endeavors might begin to build a socially liberating public standard of living.

But a full blossoming of a public interest will require a new popular imagination and vision. Building a new rationality and popular interest for a full development in the variety and number of associations, facilities, and activities will best occur on a reconstructed public rationality with a shift of popular attitudes. It will depend no less upon achieving a universal urban efficiency through creating an effective human scale for a good living environment, thus effectively substituting better living for the current glutton-defined runaway economy. So physical change and social attitudes and social behavior are necessarily interwoven. But, especially at the outset, new visions and organized public-interest efforts must take the lead.

Society as a whole must soon recognize that economic growth can be as mischievous and destructive as it is beneficial. We will have to face squarely the ponderous overburden already in cities that must be faced squarely to even demonstrate the socially damaging excesses and ecological losses. To do so, *two major realizations must be grasped, first, how society is destroying the very foundations upon which it stands and, second, how there is but one standard of progress, the human measure.* Neither of these two realizations is possible if society continues to so singularly promote profits of business and the continued growth of the GDP. The dollar measure of our existence is leading to a dual disaster, the human and the ecological. For even survival, it will be necessary to focus our values to build a new humanity—paradoxically, that is, being forced to be more human to become ecologically more sustainable. That is the "bottom line" in achieving a public-interest system.

During World War II and on the day before he died, President Roosevelt wrote, "If civilization is to survive we must cultivate the science of human relationships." Now we face a new and yet closely related need to rebuild ourselves as human beings if we are to face the twin challenges of our time, the economic domination of our behavior and the economic destruction of the foundations of our earthly ecology.

CHAPTER IX

GOALMAKING FOR SOCIAL HARMONY

Constitution of Change

If society is an instrument to expand the living possibilities of people, it follows that methods must be developed that will keep the course of society open to the best possibilities for rich and highly diverse human interests. This possibility is new, yet also a logical follow-up to the expansion of productive wealth that prevailed for the last two centuries, which in turn was a follow-up upon previous history when society was overwhelmingly organized for the benefits of an extremely small social elite. Effective social development will now increasingly depend upon developing a system of sophisticated social goals articulating the best human possibilities and assuring an adequate response to people's own growth.

Today a sophistication of goals is especially necessary to supplant the vacuous goal substitute of promoting endless economic growth leading to the corrupt runaway economy. For all of the sophistication of methods, endless economic growth reflects an antiquated misapplication of old conditions of scarcity, yet are exceptionally profitable to corporations. But they are now sending us on a suicidal course compelled by blindly promoting those profits.

Specifically human, consciously developed, and publicly approved goals are necessary to replace the current promotion of an infinite growth of the GDP, essentially a non-goal that indeed serves corporate economic power. That objective of raising the standard of living was valid until affluence was reached, but affluence played into the hands of the corporations who sought economic power through a complete control over the vast forces of social development. That monolithic and now errant non-goal has given entrepreneurs license to exhaust society and nature through compelled out-of-control competition that in reality now promotes a scorched earth for the seemingly good objective of increasing production and consumption. But now the growing GDP is becoming largely a measure of social and ecological burnout.

Given the saccharin but crushing powers of this paradoxical trap of modernity, the market system overwhelms human dignity by pressuring everyone to expand both the production and consumption. We are forced to conclude that the mechanisms organizing change—new technologies compounded by corporate initiatives—are increasingly antagonistic to legitimate human purposes. Arising out of historic scarcity, and by forcefully extending the one-time legitimate goal to overcome general poverty, capitalism has been able to actually increase its monopolistic hold upon public purposes so evident by the forces propelling the runaway economy. Such overshooting the goal of eliminating scarcity cannot be a continuing legitimate goal of a rational society. Yet today the course of change follows closely the growing corporate ability to promote spurious consumption for illegitimate human and ecological consequences. So this singular objective has effectively denied a public evolution of social development that might have balanced economic and social progress.

Therefore, to assure a fidelity to human purposes, society requires a system of specific, articulate, and comprehensive social goals to serve the widest human interests and remove us from this current historic trap of economic growth. A sophisticated and continuously progressive system of creating and evolving social goals are required, if nothing more, to keep the eyes of society on its rightful *human* purposes and become relieved of an engine that now burns human and ecological possibilities. Goals are promises by society for its citizens to help them achieve their highest aspirations. They are promises that, if reasonable headway is not made, should be called into question, altered, or abandoned, unlike today's uncritical mindless pursuit of more consumption for more profits. Goals may thus be thought of as civil rights, which in time some should become.

As Robert Heilbroner argues, society must take "its very history into its own hands." That objective has yet to be recognized, let alone be engaged as public responsibility. Consciously developed, specifically human goals must replace the unending pursuit of dollar growth and economic power. *A proper system of goalmaking is likely to become the best method of giving society decisive direction to its own historic development.* Now the power to endlessly pursue economic growth rests almost solely in the hands of corporate economics. So if goals can become a directive force in society, they must be pursued as diligently and forthrightly as the political powers established in the Constitution. Goals need to become as sophisticated as the methods of technology, accounting, and management. But, society must first wrest control of economic power from corporate hands to effectively support public goalmaking, if society is to take its history into its own hands.

American society derived its present constitution based on the assumption of an essentially static social condition in which citizen protection was believed to be

timeless. Neither science, technology, economics, nor corporations, our dominant sources of social transformation, were addressed. Social change of that time
was considered to be the development of human rights, then incorporated in the
Bill of Rights that resulted from the advanced thinking of the Enlightenment
then in full blossom. The age of steam power setting off our revolutionary course
of development was yet decades in the future. So the entire system that dominates
human affairs today exists outside the thinking in the Constitutional
Convention.

And thus the entire modern system of social transformation we live by today is
absent in the original structure of our political economy. While regulatory agencies and laws protect such things as health and safety, the *direct impacts* of products and processes, the laws we have today do not affect the *accumulative or
resulting impacts* of change, the greater outcomes of industrialization, urbanization, product development, and marketing, that is, the resulting effects of the
mass, complexity, and power of our unprecedented modern condition. For it is in
the structure of our radically changed conditions that the melancholy, disenchantment, and outright defeats and crises of our time are lodged. We have a
massive force promoting change but little to direct that change for positive outcomes. The system of change has become the objective of change, the people
being essentially the medium of the transformation.

If change is now the consistent force of society, if society is to be directed to
optimize human benefits, and if people are to benefit from the greater freedom
the modern system can attain, then higher human values must be incorporated in
the processes of our social dynamism. We are part of the greatest leap civilization
has every achieved, and so our human values should be set into the guidance
processes a proper civilization must follow. That imperative, I believe, is that the
justification of social goals effectively becomes the constitution of change.

Goalmaking, for the most part, shifts emphasis from the promotional side of
economics—blind and endless growth—to a consumer and social definition of
change, that is, to a concept of plenitude, to be addressed shortly. The process
constitutes something of a counter-revolution to the dominant forces still working on assumptions no longer valid that appeared in the early Industrial
Revolution when the objective was simply to overcome scarcity. But today, goals
made by and for people must assuredly be human in purpose, content, method,
and especially control. No less assuredly, goals must take full account of all modern existence: industry, cities, corporations, mass media, advertising, instant communication, bureaucracy, science. Goalmaking, I would propose, therefore, could
possibly close the greatest "loophole" found in the Constitution, today's immense
and unlimited power of capitalism.

A sophistication of goalmaking requires that a broad social intelligence evolve from a growing public knowledge and dialogue responding to a clear base of human purposes. That means a vigorous expansion of applied social research. It means an active, diverse, and continuing evolution of goalmaking by individuals, all communities and organizations within the public interest and all levels of government. This society-wide process would serve two major purposes. First, the goalmaking would serve each person and organization, especially to give direction to their own behaviors. Second, each social entity would contribute directly to a dialogue of society and an outreach for human growth. Goalmaking thus becomes a focal point and rational foundation for the knowledge society. Most certainly, through its careful development of process, *goalmaking can become society's constitution of change.*

Vital Goals

The goal creation process thus incorporates advanced social and technical knowledge, evaluations of experience in all sectors and at all levels of society, the active communication of human visions and aspirations, and, of course, public debates leading to various forms of goal formation, adoption, and action. Each goal-formulating person or organization acts upon its own goals and as an agency promoting (or opposing) goals at higher levels of authority. Since, however, the goal objectives serve only human purposes, commercial corporations must be limited to the powers necessary to fulfill their productive or service roles.

The dialogue, then, can find endless expressions in all levels of society, which also becomes an inspiring mechanism to understand the greater human possibilities of life. This is critical because the fundamental force of change will exist directly with the people, no longer merely the capital and initiatives of capitalism. As progress occurs, goals can find expression, be diversely debated and evaluated, and build a new level of public intelligence—a fundamental condition of an effective democracy. The process can become increasingly sophisticated in interpreting social values and human purposes. Then, too, it will advance as current progress is evaluated. The public dialogue surrounding goal formation is a forum for a valid competition among the ends and means that govern our lives. Yet most of all, it will be a system elevating social intelligence throughout society.

Where goals arise is unimportant. What they do and how they are considered, adopted, and implemented is critical. But the formation of goals should always be an active process, expanding upon and altering already adopted goals as experience and aspirations indicate. I do not imply that goalmaking should be chaotic, only that it will be necessarily complex. It stands in obvious need of philosophic and methodological foundations. Moreover, the process will challenge the theoretical basis upon which change in society occurs and to which it is directed; so there is a great need to incorporate a theoretical grounding into the roots of the social sciences and applied social research; currently the weakest sphere of science, but potentially the most sophisticated and vital.

Goal Perspectives and Discipline

If goals are to be sophisticated and reliable guidelines for society, and replace the social confinements new embedded in the endless growth of the GDP, they will require the broadest possible vision of human life in society. Most critical are philosophic perspectives. Such philosophy might become in effect a grand preamble for goals, an imperative that arises from the bare fact that only those things that are conceived and organized in society have a basis to exist in society. Such a preamble will be essential in evaluating all possibilities and contingencies, posing the assumptions, values, purposes, and qualifications that can possibly arise from activated goals. These preambulary elements, therefore, might form bridges between democracy and the current conditions of society and perhaps establish a basis upon which adopted goals might later be judged. From those platforms, both very broad and very specialized goals may be debated, adopted, and set into motion.

Since today's monolithic "goal" of promoting only endless economic growth is sought at a sacrificial price of social crises and ecological devastation, many early goals must necessarily be directed to their amelioration. However, it is critical to put goalmaking on a positive plane and, where possible, include the social amelioration as a part of very positive human objectives. Thus many social issues of family breakdown, human neuroses, or crime should best be rationally a part of building an unequivocally healthy society. Goals for the development of a universal urban and social efficiency, for example, can directly reduce both human stress and ecological debilitation while seeking imaginative social environments. The most positive human aspirations must prevail by always keeping them in the forefront of social thought.

Goalmaking is a means to establish a fidelity between the organization and operation of society and the human values they serve during the complex social transformations. Had this existed over the last century, we might have discovered, for example, that the objective of urban form is to minimize, not to promote the growth of urban travel, among other urban goals. If we had fully examined our aims and their underlying assumptions, or had we projected the consequences of making the city over for the incredibly inefficient automobile, we would have promoted rather an optimum system of access (mainly through urban design), and avoided human domination by mobility and the massively destructive urban sprawl associated with it.

Properly developed, goalmaking imposes a new discipline of thought bearing upon the objectives and methods of public action. Had we avoided the disaster of sprawl and excessive (and yet never fulfilled) mobility, we would have built the city to the human scale, which means building a humane urban environment. We might then also have been led to discover the many ecological, social, and cultural advantages of urban form (perhaps without fully realizing or even needing to know that we had done so). Most likely, we would have also evolved forms of human community. As it is, the still small ecology movement at century's end has not grasped the dynamics of urban geometry, let alone the closely related social benefits. No less, an awakening offers only a partial vision of some of the great defeats of modern economically generated disaster. Even today a vigorous dialogue built around human goals could be a match igniting such a cluster of powerful human and ecological insights.

If goalmaking is ever to measure up as a constitution of change, moderating our all-too dynamic times, and setting forth a broad scope of social democracy, its validity will likely rest upon six foundations: (a) a broad base of continuing applied social research, (b) the development of varied philosophic orientations (including competitive themes), (c) the varied development of candidate goals arising from many sources, (d) an extensive dialogue about the merits of proposed goals for whatever level of society they are intended, (e) a democratic process of adoption and program development, and (f) a long-term program of review, evaluation, and evolution of the process.

If goalmaking stands as a constitution of change in a dynamic society, it should be viewed for its supreme importance in maintaining a clear and prevailing social fidelity of human purpose. Only a penetrating perspective at the constitutional level can help us know that human change is not made simply upon products and material wealth or upon blind forces of corporate economics or any other kind. Considered from the viewpoint of guiding fundamental social change, goalmaking therefore inevitably takes on a magisterial quality, like the Constitution itself.

We might glance back to the central terms of the French Revolution over two centuries ago. Its ringing concepts of *Liberty, Equality,* and *Fraternity* today tell us how little has been accomplished since 1789. The greatest progress has been the vision of Liberty. Yet, even that concept remains mainly a restraint upon governmental power, and is not considered for *higher positive social opportunities* as an objective of society. While material progress has occurred, we remain democratically backward in directing economics to its full human Equality *a system of united national production but without a united system of consumption.* And if we examine Fraternity, we know that society has fallen into a fragmented, alienated, and pulverized condition of economically exploited social masses, not one of a *socially coherent, personally secure form, such as community.* Certainly, Liberty, Equality, and Fraternity are among the most daring human aspirations to echo across the human consciousness. One might suppose, however, that if those words had been translated into articulated goals and plans throughout the last two centuries, actual human achievements might have been far greater.

National Urban Accounts

Cities guided by a goalmaking process helps to propel them into becoming a chief organizing system for society, especially because cities can promote development of a rising *public* standard of living. Any institution of such an importance to society must be measured precisely for all of its effects upon its citizens if that complex organizational framework is to most effectively serve those citizens. A valid accounting must thoroughly cover both the cities shortcomings and benefits, evaluating their vast array of livability factors.

Until now cities have been known mostly for their problems while stagnating in continuing crises we now consider normal features of life. They exist without a vision of urban progress, attesting to their low esteem and their history of being commercially exploited by the real estate market combined with their absolute domination by automobiles. Hence, a process of goalmaking served by comprehensive measures of living qualities will be central to reversing this sordid history and establish cities that will be in the forefront of a new quality and character of progress in the modern organization of life.

The most relevant model to measure urban progress is the national economic accounts that measure the seriously flawed gross domestic product that now reflects the misdirected growth of the runaway economy and its dismal per capita income counted as the standard of living. Yet the intent and scope of urban accounts can reflect the precision we need if urban living is to become valid as a major organizer of society—as they must if humanity is to flourish. Such urban accounts will be inevitably complex if they are to render a reasonably simple but valid measure of good urban living, while stimulating social insights for progressive aspirations of urban living.

National urban accounts must address the broad range of qualities of urban living that society can aspire to. That means natural ecology and highly congenial human physical environments. Both measures can help promote an unprecedented universal efficiency and the range and accessibility of recreational and cultural facilities. Communities and community buildings offer the best social and structural conditions for an optimum human participation and an effective urban geometry—serving people at least as well as we now organize production. Minimizing automobile use through urban design will emphasize the great efficiencies of walking when environments are built at the human scale and superior transit when based upon concentrated developments like communities.

Confronting social problems must be included in any valid index of urban quality. These conditions include youth delinquencies, turbulent family life, an environment of street crime, a drug culture, organized theft, gang warfare, and violent crime. These behaviors largely reflect poverty, poor education, lack of job opportunities, unwholesome neighborhoods, learned intergenerational anti-social behavior, and racial biases. Thus both the debilitations and their background conditions are important measures of low urban quality.

Positive qualities of urban life can be reflected in healthy civil society, including high levels of public dialogue and the goalmaking process itself, the growth of traditions and vigor of social festivities, the variety and energy of community organizations, and the kinds and qualities of civic leadership. Involvement in the arts, local philanthropy, and the level of citizen participation are significant. Important, too, are programs directed to resolve and avoid social problems and to pursue local visions of good living. The list of useful measures can be expanded; all need to be assigned weighted values for indexing.

Although community is not itself a measure for urban accounts, it includes a comprehensive range of features that coincide with the highest qualities of urban life. As a composite, integrative structure of urban living, community is a force for human benefit like a corporation is a force for effective economic performance.

The range of factors to be included in an index of urban quality are highly varied, complex, and intellectually challenging; but they can perform like an instrument panel guiding the course of social development, indicating our social position at any point in time. Then they can be effective bastions for debate and continued goalmaking. The measures should be accompanied by written evaluations that qualify values, highlight important features, project trends, reveal opportunities, and make proposals for altered programs, making the index itself a highly valuable tool for constructive change. And as a socially anticipated report, citizens can be invigorated to understand and support the process of positive, publicly initiated change. And being as truly diverse as human motives themselves, the particular measurements, the various grouped indexes, and the final combined index will be socially beneficial far beyond today's monolithic GDP, which once usefully served society during the rise of the standard of living but has now become destructive through the confused support for the runaway economy so tragically demonstrates. While calling the index a measure of a new public standard of living may be initially appropriate, even that standard of living should one day give way to higher, more articulate social visions.

Working Ideology

In the modern system involving great social mass, complexity, and power, our condition today requires something more than sophisticated management.

Assuring social purpose is critical if the mass, complexity, and power are not to distort or destroy human essence. A new ideology is necessary that formulates human purposes for everyone to understand and direct, to identify with, and establish a new social dynamic. Both capitalism and democracy represent well-seasoned ideologies with endless variations. These systems were tested severely in the Great Depression, two massive wars, and throughout the long Cold War—history's greatest test of competing ideologies. Capitalism and democracy also largely define American patriotism and strive to formulate life's satisfactions as well. They set forth how social power is achieved and applied, political power on one side and economic power on the other. Indeed, these powers constitute the two great and interlocking games continuously played out in American society.

However, there is a simmering discontent concerning both of these ideologies, not from any obvious failure of either, but rather from their successes, excesses, and side effects in economics and simple ennui in politics. The great question concerning the malaise of democracy and disequilibrium of economics is that they reflect the need for a new ideological framework, one that can give people an active and vital role in social affairs. Such an ideology can stimulate new visions of life, might point to new, more human forms of progress, and set forth a new quality of human association. Such an ideology would complement those of democracy and capitalism—and possibly qualifies them in human terms—especially to moderate excessive economic powers. Ideologies don't just exist. They are conceived, developed, mature, and change with time. In its own way an ideology of the *public interest* must achieve an allegiance—no less than democracy or capitalism—and build a human esprit. It is precisely in the realm of human esprit that is most vital today in forming an ideology of the public interest.

To a very large degree, a clear public-interest ideology can reflect and amplify the fruition of both democracy and capitalism, giving them an increased strength at their core, for in their present forms both have, we might say, seriously run their course. This is especially true for economics. The sheer ability to produce—which could also be readily expanded—directly reflects our historically new power to draw down resources and despoil so much of nature. The growth of environmental excesses may be associated with the fact that there has not been an effective "counter force" to give it a new and positive social vision, to offer alternative plans, and to apply restraint where necessary. These are actions that the environmental movement has sought to apply, but so far with only limited success. The movement is too easily seen by society as a sentimental upstart, a contradiction to the greater "realities" of the market place, that is, without historic authority, operating validity, or continuous initiative. In other words the environmental movement is in need of a larger framework, that is, a more comprehensive ideology, uniting human ecology into the larger scope of humanity and society.

The case with democracy is different, almost a contrary example, in that the primary public mood is repelling disgust. Michael Sandel has expressed in *Democracy's Discontent* how "individually and collectively, we are losing control of the forces that govern our lives." Despite our wealth and a still reasonably functioning democracy, there is, as other authors have noted, a deep sense of powerlessness evident in the affairs of society, both collectively and individually. Human beings require a new foundation of creative challenges that not only resolve their powerlessness but, more significantly, give greater verve, value and enduring meaning in all avenues of life.

A subtle but critical disconnect seems to be occurring among persons and in their identity with the economic and political forces operating in society. The person, though often affluent, cannot be associated with any public goal in the larger society or for the self, except in producing and consuming. So earning more money and watching television are the prevailing outcomes. An omen of disenchantment pervades active living, whether founded on reports of environmental destruction or an urban society subsiding into meaningless morass of frustrating mass functions. Anyone who bothers to keep up even marginally with public affairs is unlikely to find inspiration to build higher motives. It is a good time to be a woman, to be sure, as more move into positions of economic power. But, for men, they are not even maintaining equal registration in the universities and colleges.

There are no vital end goals to stimulate serious motivation or a clear personal identity, beyond exerting class-defined expenditures. There is a decreasing satisfaction to be found in merely expanding gross consumption, as Galbraith noted way back in 1958. With assurance, we can say that the pleasures of consumption, while everywhere welcome up to the appearance of affluence, thereafter have a diminishing proportional value as wealth increases; for a first new home, the pleasures of comfort are great; but they don't rise proportionately when the family can afford a mansion; then they merely demonstrative an ego of class luxury.

But increased personal wealth does become a very attractive target for commercial promotions. For all of its power of influencing buyers, advertising basically promotes an invisible cynicism, as much for those who don't think about it as for those who do. That is largely because the days of introducing new basic kinds of conveniences are over. Cell phones and the Internet do not compare with refrigerators, air conditioning, or even automobiles. No great social objects are in view, now that the frontier, growth of industry, war and cold war, has faded in the social mind. Growth, and only growth, remains the order of the day, and it offers basically little new except a relentless reign of novelty for the hand or mind.

Can a new ideology built around the public interest help resolve the growth of ennui? Certainly it may bring people to think anew about their lives with others, of associating and cooperating for what their imagination inspires. In addition, each community, as an independent unit of the public interest, holds the possibility of creating its own separate ideology of living, diverse and challenging personal and interpersonal pursuits. Community can be formed as an ideal for every person to be stimulated and pursue short term or lifelong goals, most of all including those of their own imagination, thus returning the focus of society to the people in their living environment. A *personal career* alongside a *professional career* effectively removes persons from a total dependence upon their work life for their singular socio-cultural identity on one side and choices of products and services for their personal satisfactions on the other. For many persons the source of producing and the outlets of consumption no longer constitute a worthy self-empowerment.

Then, most significantly, embedded in the core of community behavior is the concept of personal self-development, a process in which persons can set forth their own life-long course of promoting unique and personally meaningful experience. The concept has endless possibilities and they point to a human growth that accelerates with the growing diversities of civilization. The most significant progress by individual persons in recent history is one of increasing the abilities of persons to pursue interests independently on their own behalf. Yet this human capacity has still to be fully developed in personal terms. Self-development within community both formalizes and individualizes close association to promote, first, socially healthy persons and, second, independently vigorous persons pursuing life's interests to their greatest meaning.

Aware persons have long said to themselves, "When my financial goals are realized, not only can I live in greater comfort, I will then change my life." The problem is that society sets forth the means for material achievements, but the person is then left completely on his or her own devices to make the change in life, without training or preparation, without either a concept or opportunities of self-development. Here again is a demonstration of the fact that little exists in modern society that is not planned and organized. Community changes that, first, by providing the physical and social environment for opportunities and maximum ease for self-expression; second, by setting forth the concept of self-development and in time providing an increasing number and variety of creative role models; third, by offering diverse learning processes; and, fourth, by offering all persons direct assistance and endless options for their own interests to grow.

With today's achievement of great material wealth, the singular system of the market place of purchased satisfactions loses much of its appeal. Turning to each person for self-growth, the community gives the life of the person its own unique spin. When assisted by the community, the person can grow in his or her own terms. So community, through self-development, directly shifts the locus of social action to the person, one might say, as a part of civilization's service to the higher development of human nature.

Plenitude

As deeply as I have opposed the *endless* pursuit of wealth, already in some quarters that singular pursuit of wealth has wilted. What is necessary is not merely a negation of the old motivation but rather a new human ambition that in time will possibly to supplant the possessory motivations. And, for millions of people, there is a already a partial negation of the old force, replaced for a large population by what the authors, Paul Ray and Sherry Ruth Anderson, call their book, *The Cultural Creatives.* The authors present evidence that there are fifty million such people in the U.S., most of whom appeared after 1960, people who have established a new awareness of life's greater possibilities. Those persons are associated with a number of key values: globalism, the environment, civil rights, women's new awareness, and new levels of altruism, self-actualization, and spirituality. Altogether they signal a mind shift in the way they think about themselves and society. They do not constitute a recognizable overall movement. But they do reflect, I believe, a readiness of an already large population for a massive shift in the mind of society.

In the 1960s Lewis Mumford predicted such a movement, concluding his *The Pentagon of Power,* saying "we shall find that the necessary change of attitude and purpose has been going on beneath the surface during the last century, and the long buried seeds of a richer human culture are now ready to strike root and grow, as soon as the ice breaks up and the sun reaches them." (pp. 433-434) So perhaps the Cultural Creatives are striking that root and growing as yet in separate, still largely inchoate attitudes and remain without a clear vision of broader possibilities. In any case the combination of a slowly growing demise of the money motive and disenchantment with raw promoted consumption, combined with the ecological and social issues, is having a certain effect. Already, too, there are indications of a growing desire for self-development.

Yet, much of the developing disenchantment arising around wealth has to do with how wealth is used. A movement for more personal development inevitably revolves around the notion of plenitude, that is, how much wealth is ample to underwrite alternative, essentially non-economic orientations of life. Still, the level of economic attention given to conditions of the "enabling" side of consumption has not been a primary focus for economists, unlike the direct promotion of consumption. But, as I have stressed, the forces behind the runaway economy reveal that pump-priming to raise consumption has become the major force promoting economic growth, a fact that the Cultural Creatives must feel in their bones if not seen in front of their eyes

Consequently, if there is a shift of mood away from raw growth of consumption in favor of specific human interests, then we are in need of a principle to evaluate one's personal resources, including money, time, and other assets for good living. Lewis Mumford seems to have edged very close to a new concept with his repeated use of the term "plenitude," the condition of being ample or complete. Such a term suggests that there is a basic consumption that sets forth a saturation of classic consumption beyond which the *qualities* and *benefits* of the good life are the guiding terms of plenitude.

This kind of distinction contrasts sharply against the traditional unlimited private spending for consumption. The advantages implicit in plenitude are numerous. Public goods serving everyone are far less costly and are less disruptive of behavior than the same private goods privately purchased. In other words, two or three large community swimming pools cost much less and would be used far more than twenty or forty private backyard swimming pools, plus the facts that everyone "owns" and uses the community pools and this brings people together socially on equal terms for collective recreation, thus underwriting and enlivening public events. An understanding of plenitude, I believe, thus sets the stage for varied public involvements and a role for the public interest.

Common and multiple public uses of spaces and facilities are also far more important than merely their savings would suggest. The greater value is how they assist people to rub shoulders, enjoy association, create common understandings, and prompt many forms of cooperation. Multi-use facilities, such as pools, gyms, schools, and studios, by uniting them in a single design at the human scale, afford a vigorous social life. The human scale greatly eases association between people, helps unite them by involving them in varied activities, and supports an ease of casual association throughout their comings and goings, perhaps a short visit, possibly coffee, but always a greeting.

When developed, plenitude becomes a strategy for full living. At its core plenitude is a *living efficiency* and *strategy* that conserves and optimizes all elements of personal achievement. Benjamin Franklin anticipated the concept when he claimed that a person could maintain a full life (in the terms of the eighteenth century) while working only a portion of the time then the custom. He thus implied that leisure and its uses were the higher object of work, not an infinite growth of wealth for gluttony. Work in Franklin's mind was only one part of a strategy for full living.

Being a tool in the strategy of articulating optimum possibilities of each person's life, plenitude grows especially upon the expansion of universal efficiency—and the leisure or leisure activities that efficiency affords—and a conservation of money, time, space, nature, and resources. A personal efficiency especially permits a shift of life's resources to the qualities of living. And it is community upon which the good life turns, first in making possible personal efficiency and, second, by offering the facilities and programs available in the activities of good living can emerge. No less important, community overcomes our current deficit of meaningful association, making interpersonal behaviors a hundred ways easier, thus setting the foundations for cooperative behaviors with the community's abundant facilities and responsive programs. Finally, community can overcome today's depleted cultural involvement, now effectively limited to wealthy patrons of the arts. Hence a high level of plenitude is generated through community offers an efficiency and conservation of all ingredients of good living.

Plenitude thus involves two elements, a private consumption with conveniences, and an optimum level of association and opportunities for personal self-development. Plenitude sets an abundant life-fulfilling potential based on ample sustenance and auspicious potential.

Plenitude is like the community that gives it human force and magnanimous diversity, a principle of human empowerment, a human redirection of today's increasingly wasteful growth of rude production and compelled consumption. We can measure it in terms of its efficiencies: personal production and consumption, time available for leisure activities, uses and conservation of material resources, broad availability of facilities and supportive programs, the near instantaneous and universal access and minimal need for mechanical transportation, the ease of interpersonal association and cooperation, and ample capital for personal sustenance and amenities. And, based upon effective goal making, these measures can be effectively substituted for the now disenchanting and threatening growth of the corporate economy and the increasingly consumptive standard of living.

Economic and Social Justice

The first concern for a public-interest system is assuring for everyone a guarantee that there shall be no poverty anywhere in a wealthy society. That objective has been economically feasible throughout most of the twentieth century. Indeed, it was all but politically accomplished in the early 1970s when a guaranteed income passed the House of Representatives and nearly did so in the Senate, but died at the close of the session. Instead, economic growth continued to increase, but increasingly centered on expanding the runaway economy.

Poverty in an advanced industrial system is not merely wrong morally; it is illogical, for the economy cries out to fully utilize its productive capacity, as it now does by promoting the runaway economy. Similarly, poverty in a democracy is socially absurd, if not cruel, and is an historic embarrassment in the same character as slavery, where people through common economic output are bound to it, although it is not bound to them. Since the productive system is nationally unified, so too should the distribution of wealth.

Two other economic issues require attention. Universal health care has been proposed frequently since the administration of Harry Truman about 1950. But today, despite having the most advanced technical medicine, the United States falls low among industrialized countries in the various measures of health and health service, especially for new mothers and infants. But the power of proprietary health care, the pharmaceutical industry, health insurance, hospitals, and doctors have over the decades invoked what I have called the Marxian pall, in this case branding a national health service as "socialized medicine," a meaningless but lethal political condemnation as a step into communism.

Second, public education has been in continual shambles for most of a century, moving from crisis to crisis and following many failed "renaissance" movements. Commitments to true excellence in learning are bogged down in a century of increasing bureaucratic overhead, a burdensome tradition of a class-teacher-grade structure, bound in the conservative tradition of school boards, and beset by overbuilt and bogus teacher credentials. Little can be done within the present system. To stimulate true reform, it is imperative to break up the present school operations in all of its basic structures: one teacher standing in front of a class; presenting uniform class curricula; giving grades for conformity to engorged knowledge; and confining learning to given hours, days, and months—all preventing creative personal growth. With the breakup, the entire purpose and methodology of learning will be exposed to and require original thinking.

We must honor diversity of learning if we are to honor the diversity and liberty of persons. That means stimulating and honoring individual creativity in learning more than any rote accumulation of knowledge. Little wonder that there is a nearly universal student depression, a depression preparing persons for the production-consumption roles in life and denying them imaginative living. Progress will undoubtedly have to await other broad social reforms, of which learning must then become a part. Community and its commitment to self-development await such a possibility.

Two of the greatest focal points for a public-interest ideology centers upon the two basic forces promoting the runaway economy. Urban reformation is at once the most grievous issue and most promising form of human renewal, grievous because it is buried in pipes, hard surfaces, and costly structures, promising because it will undoubtedly be central in founding a modern renaissance that gets down to the roots of what life in society can be about.

The second focal point is that the advertising dominated mass media represents possibly the greatest propaganda system ever devised, manipulating the most universal communication system yet created, and reducing the mass media to control the minds of people by reducing messages to their lowest common denominator. It simply pollutes the right of a "free" press. The challenge is to promote the public interest to effectively wrestle with the cluster of critical issues of open communication in a free society. Certainly the right of a free press cannot be construed as a license to reach into the human psyche and shape the mind of society merely to advance sales. Thus a public-interest system may find its greatest challenge to reorganize how dialogue and social communications now take place.

These matters are but a few of the elements for the public-interest ideology to take up. The broader questions strike to the heart of what the course of social transformation is all about. As presently organized, society is not positioned to deal with the profound questions of what form the future society should be. But the many issues are there, and in one fashion or another, hopefully one of intelligence, they are urgent questions that must be confronted.

Integrative Harmony

A central concern of the public interest is to guide social change to serve the highest human values as people themselves determine. Once the general economy reaches the level of affluence where a freedom of expenditure becomes significant, then the evolution of society importantly shifts from its technical and economic foundations to the social and cultural realms. The technologies and economic forces will then recede in society, even as they provide an increasing support for the socially determined objectives. At that point the public interest takes on the chief role of determining the course of social change in society. From that time the social evolution will be propelled to the forefront for a broader social accomplishment as determined by the people through their goalmaking process, that is, the public interest stresses a line of development close to the higher purposes of the civilization.

Through community and self-development, the aim is individuality in its most robust meaning and a diversity of social conditions that offers a maximum possibility for self-defining self-development. Indeed, everyone becomes his own greatest influence and elevates the dictum to know thyself into a new and spirited consciousness. Community is integrated to serve this greater human freedom. And it is by self-development that the community can demonstrate its highest accomplishments, by contrast today in which one's fractured existence in mass society is reflected in a disconnected, unfocused, and increasingly meaningless pursuit of material rewards. Today's most integrated development affecting the person is preparation and pursuit of a remunerative career. But, in community it is self-development that both unifies human behavior and yet decentralizes authority to the self-controlling person. In this way community offers a new, local, and dynamic synthesis for both freedom and order.

Hence, while democracy today remains unfinished and limited to its legal and political roots, community offers a concrete, local, and positive social growth of democracy. If, as Spinoza argued, the only true objective of government is freedom, community exists as the immediate agent of government, and positively pursues an infinite range of freedoms. In other words, *freedom becomes a social, positive, and personal pursuit and community is organized to assist that pursuit* through its many readily accessible facilities, organizations, and programs, all suited to the unique interests of individual persons.

The many possible lifestyles within a given community reflect its many programs and environments supporting the central purpose of *self-development*, reinforced by the community's own conscious self-development. Community is the most important tool and even a major object of *goalmaking* that guides social development. Because the community is constructed at the *human scale,* it can then achieve *universal efficiency* and the *leisure* for casual living and for vigorous pursuit of one's interests. The community's many voluntary organizations offer a means to expand individual *initiative* and *social ambitions.* All of these conditions promote varied and profound personal *experiences* throughout life. All can invigorate the *amateur* and the role of the *volunteer.* And when people are free in this manner and in trust with one another, they become experts in mutual *self-entertainment*—possibly the most dynamic and spontaneous diversions—in conversation, humor, music, and locally organized festivities. Community, therefore, becomes a social compact of intense mutual interaction: voluntary, casual, varied, face-to-face.

The diverse lifestyles thus reflect and promote two foundations for an ideology of the public interest. The community becomes a city made small and coherent for the person, revalidating the locality of life; it promotes a diversity that is rich and personal; the community building is constructed efficiently to expand free time and yet offer many stimulating ways to use that time; a strong partici-pative social involvement is heightened by individuality and easy options for pri-vacy. Altogether, community creates a system of fundamental human empowerment, a personal power hardly possible in the individuated mass con-formity of today's society.

A second foundation for a public-interest ideology to evolve is in the unlim-ited range in which metropolitan organizations and national, specialized, volun-tary associations as cosmopolitan counterparts to local communities can pursue the varied strategies to strengthen public life, serve their members socially and culturally, and promote continuing dialogue on the most basic public issues and great social possibilities.

A new stature and power of the person is implicit in community, especially in self-development, in providing persons access to the privileges and qualities of its public life, all in addition to today's personal consumer privileges. Self-develop-ment can become a right, the right to individuality—as opposed to suffering individuation in mass society. One cannot claim a right for spontaneity and indi-viduality, but community might assure the conditions upon which both may be freely expressed, at least giving them a human legitimacy now suppressed in cyn-ical depression.

Democracy has yet to come to terms with many critical elements of modern-ization in society, including: (a) the technical world, (b) the mass society, (c) urban development, (d) the corporate grip upon human sovereignty, and (e) the advertising-driven mass media. The public interest—a social movement more than any particular endeavor—cannot be expected to resolve these vital issues, but it can frontally address them, always keeping the person in focus, and possi-bly making headway for some of them over the decades. The important matter today is that it can offer largely new tools to strengthen the human content of democracy.

The greatest potential for the new sector of the public interest lies with the creation of viable communities to serve the person directly in the most humane manner. That potential must be both created and learned, and initial work to cre-ate modern community may be forced to compromise with some of its principles. However, compromise has no role to play at initial conception because *the first essential matter is to convey the fullest potential of community* with as much clarity and integrity as possible.

The broad aim, first and always, is the development of a socially cohesive and congenial human tradition. Perhaps we can at least ask whether such a human prize might become a significant tradition of civilization.

New Avenues for Discovery

In practical terms, the community's role in structuring human environments for a broad efficiency will mean that society will become as aware and diligent about consumer costs as producers are about production costs. Then people will be able to organize their lives around a complete package of efficiency, ecology, amenity, and immense social advantage. And they will do so without the severe constraints of time and money that dominate behavior today. We are a society built around time and money, so those values dominate our thinking and become very real obstacles today when people are struggling, for example, to put together a proposal to put on a play. They typically must find a vacant facility and furnish it in an ad hoc fashion. All costs must be born anew for their specialized use in an isolated, ill-fitting urban facility that makes such enterprises into severe hardships, and then abandon it after one or two plays are presented. Years later a similar effort must start from scratch.

Our cities are now a world of miscellaneous conditions set in miscellaneous circumstances for socially miscellaneous outcomes. This is especially true for efforts to support the arts. While business enterprises compete and profit with this condition such approaches to human purposes is inherently inefficient, socially incoherent, and, I must say, basically inhuman. By contrast, communities with known populations, known product and service needs, and known ranges of interests need not suffer by it. Needs are measurable and there is no need to waste by preparing for too much or too little. The few things that are not known could be accommodated by measured reserve levels of spaces, whether for utilities or facilities.

Society, having fallen to economic determination, means that social incoherence is the normal and expected conditions of life in society. But since that incoherence is associated with the great growth of wealth and a wide choice of products, it is assumed to be right and good. And, because the newly won affluence has permitted the corporations to manipulate the consumer's side of the classic market, the social chaos directly serves that marketing process. Then the corporate ability to force-feed economic growth takes on a character of public bribery.

When society can constructively face up to the blind consumption that arose with the appearance of general wealth, reorganize the logistics of life; we can support an abundant plenitude and a socially expansive direction of change. Then we will be able to effectively disengage ourselves from economic domination and rise above the confines of mere exchange-based affluence.

Community may be considered as a generator of human opportunities, beginning with simple leisure and comfortable amenities arising from a plenitude with new, positive, and social freedom. With economic security and personal empowerment, people can begin to feel that society is acting completely in their favor, not merely remain disadvantaged activists in the market place. Julian Huxley foresaw the possibility of a new economic freedom, writing during World War II his small book, *On Living In A Revolution,* suggesting that "democracy must extend into the economic and social and all other aspects of life if it is to be complete." Huxley emphasized how individuals are "the yardstick by which we can measure democratic method" and stressed their "active and voluntary participation in all kinds of activities." These perspectives are bound with the communities' ability to expand the personal and social freedom, bound, that is, close to the achievement of universal efficiency and the possibility of plenitude, with a restraint upon economics that opens new avenues for the advancement of social democracy based on the growth of personal opportunities.

Community is built to the particular qualities of human nature, specifically, on a flexible format for people to evolve as they will to their own best outcome. Many discoveries and creations remain to be made to serve that purpose. Some include, (a) a highly dynamic geometry of urban space; (b) the social vitality arising with the human scale and integrative urban design; (c) a person's time and money unburdened by incessant mobility and compelled work; (d) a human ecology that is fundamentally harmonious; (e) a second and specifically human career not monopolized by money; (f) the discovery of life-long and continually growing friendships; (g) a greater richness and range of human experience; (h) development of personal identities and traditions; and (i) the endless variations of self-development.

These avenues of discovery can be multiplied and be realized because communities are structured to address these interests and widen horizons of human progress. Their design is organized for human diversity, whereas today's market diversity is fundamentally disintegrating in social terms. The limited size of community is closely related to its efficiencies and freedoms. The human scale of both environments and institutions is the community's most universal feature, a principle supporting major human possibilities. Rousseau argued—even in the small urban scale of the eighteenth century—that community should be "proportionate to the limits of the human faculties." He stressed in *The Social Contract* how "seeing and knowing one another should make the love of country rather a love of the citizens than of its soil." Scale was thus also central for Rousseau through which other qualities of his community were to be derived.

If the concept of community has merit—a principle of balance that organizes social affairs for the wider horizons of living—it is because community represents rational limits on all things in support of wider human objectives. It must achieve a clarity such as the scientific method became during the Enlightenment. It must be backed by a continuous improvement, like technology after steam demonstrated the broad inventive potential. And it must be continuously dynamic, like private enterprise became when industrial forces reached their catalytic stage in the late nineteenth century.

Human careers are bound to the conditions of their time, but they can also change when new options present themselves. No age knows this better than our own. We can be bound by our present condition, regardless of the human and ecological distress. Or we can consciously seek a new, more expectant reality, one that challenges old assumptions and today's erratic organization of life. My case rests with the potential of community, its inherent course of plenitude, vitality, and harmony.

Inevitably, the many avenues of social discovery can now signal a new, conscious, grand, and inspiring pursuit of good living. A flowering arising out of the productive system today—for the first time in history—permits a conscious growth of human freedom; a relief from the grueling psychological toil of our industrial system, like the freedom we achieved from the grueling physical toil during the early part of the Industrial Revolution. Ostensibly the better human possibilities like the arts should be high on the list of social priorities. But in our system they are unlikely to be on any list at all.

CHAPTER X

FAR LIMITS OF FREEDOM

New Instruments and New Rights

What is human life if it can't be free? What are the social conditions that limit or expand freedom? What is our ability to experience freedom and use it creatively? What are the meanings of protective and positive freedoms? Can human freedom be attained and become meaningful if it is not local and personal?

These questions, I believe, are highly relevant to the advance of freedoms that have yet to become a part of our democratic aspirations and heritage. The content of the good life has always been close to the range of freedoms people enjoy. This belief was clearly central in the Declaration of Independence where among the inalienable rights it foresaw was "The Pursuit of Happiness," in which quite evidently the growth of freedom could be anticipated. Moreover, the Declaration continued, "that to secure these rights, Governments are constituted among men," which implanted freedom in human organizations and today would have included corporations as well.

Yet, our condition today seriously belies the social intent of the Declaration despite our unprecedented material advancement. We live in a high-potency society which today offers a foundation for a huge social realm of new progress. But with an overbuilt economy, that foundation also denies its best human development. Rather, as I stress, deep social distress and ecological destruction reign, creating new but spurious foundations for economic growth, ironically a growth that feeds upon its own excesses. The political power behind economic determination also neutralizes critical political responses that might help stimulate a new public interest sector of society, particularly the development of community. The system thus effectively blocks a positive social evolution of society. The commercial vice grip overwhelms social and political thought.

Unquestionably, we need to recognize, as the world's oldest active and foremost democracy, that we now exist within a severe deficit in our vision of the human possibility. We are not as egalitarian as some other advanced countries, the only one without a comprehensive health care system, even as we pay more for health services. We have higher rates of violent crime and the highest per capita prison population. Yet our system promotes—and in cities demands—gross gluttony. This has worried many thoughtful persons for many years, like Zbigniew Brzezinski, who warned, *"Unless there is some deliberate effort to reestablish the centrality of some moral criteria, for the exercise of self-control over gratification as an end in itself, the phase of American preponderance may not last long."* (p. 33, "A world Out of Control," *Current Books,* Spring, 1993, original italics)

Our society has been economically capable of supporting a new and expansive human promise for most of the last century, even as we maintain a level of knowledge and are capable of organizing the human career on earth at a level that was difficult to imagine in 1900. We can now assure fulfilling all material needs, even with one hand tied behind our backs; achieving social conditions that eliminate all basic needs for everyone; developing secure conditions for a healthy personality; eliminating the bestial social conditions which now propels socially generated crime; eliminating the massive grinding of gears constituting the city; developing facilities and programs for expansive social participation; offer greater efficiency and more free time for everyone to reorient the terms of their living; establish environments of beauty; minimize arbitrary consumption and therefore underwrite ecological sustainability; and set a new, inspiring basis for everyone to grow, develop, and mature for a positive life in society.

These possibilities are not mere speculations. They are real and feasible with the technical and economic powers now existing. They were proved in one fashion way back in the Great Depression, which was a collapse of the ability to consume while much machinery remained idle. They were proved dramatically by the growth of production for World War II and the Cold War that in both cases substantially increased material consumption by the people. However, these possibilities exist mostly outside the market system that grows increasingly upon burdensome private consumption. Today the market exchange system overwhelms us, yet confines us, and grows upon our confinements.

So now "a more perfect union" requires its own evolutionary process: a powerful non-market process of founding a *social determination* of life. There are no inherent blockages or theoretical hurdles to stop it. Thus the promise, hardly conceived because of the Cold War ideological Marxian pall on creative thought, nevertheless stands starkly before us. As a society, moreover, we have never tried to discover the grand *social* possibilities before us, having put our store of emotional energy and intellectual effort into overcoming material scarcity. The great promises of our time, now however, are the personal-social-cultural dimensions of a new and inspiring life.

These freedoms can in time become as vital as the Bill of Rights. But they are different in many key respects. They open doors to *positive* human growth. They are *active* and *developmental,* not static like most of today's rights; the idea of human progress is central to their concept. They are *social* avenues to progress, not political and legal restraints. They are *personal,* affecting people directly in their common activities and pursuits. One may also argue that they are *visionary* in that they encourage people to perceive their lives in terms of its specific human potential. They are for everyone to select and develop in the manner they themselves define. But the potential utterly depends upon the creation of new social arrangements and organizations based upon clear purposes and direct public action.

Our possibility is nothing less than a new range of human freedom pointing to an expansive scope of human progress. While we tend to think of cities and communities in passive terms, they can become decisive instruments critically advancing new freedoms.

These freedoms may include: (a) a secure and comforting basis for healthy childhood development; (b) an evolving basis to promote personal self-development; (c) the achievement of a universal social efficiency in the built environment to attain increased leisure and emotional ease; (d) the preservation of large and varied natural areas for ready, close access; (e) a close associational involvement with other persons of like interests; (f) an articulated learning process spanning from the cradle to the grave; (g) a healthy atmosphere with immediate free access to health care; (h) the provision of numerous social opportunities for everyone; and (i) a major counterpoint between the many options of community and cosmopolitan affairs.

These freedoms await the creation of a social system of human advance, comparable to the current organized system of private enterprise. One day they might evolve to constitute an ideology of the public interest and be expressed as human rights.

Other rights must be expected, but these can keep us busy to get started. Their primary instruments of action are cities and communities. Their universal foundation, I stress again, already exists in today's economic abundance, which of course requires reorientation.

As we may now view them, these dimensions of progress might be considered as potential rights that are first set forth as social goals, but always understood as a process leading to free and continuous *social evolution,* eventually to become binding *social rights.* They are conceptually universal. Their achievement may be considered as intellectually as broad and as challenging as the historic race to production and the growth of the standard of living.

Creative Freedom

On one matter we must be especially cautious. I doubt that the typical person is psychologically prepared at this time in history for a sudden, new, and greater range of freedom, probably no more than European peasants of 1800 were prepared to engage in competitive enterprise. Similarly, neither is society able to absorb and promote new levels of freedom in huge doses, just as basic democracy has proven to be awkward and precarious for a society without democratic experience. The danger of a sudden new freedom might be compared to an especially rich diet served to a hungry populace after it has existed in a starved condition.

When a large new social arena of freedom is presented to a people, we can expect that there will be some excesses and unfortunate experiments causing back sliding toward a previous non-free society. Similarly, capable scientists are unlikely to arise from the very confined learning environment of a newly developing country. A Nobel Prize is unlikely to be given in this generation to a chemist raised in Mogadishu, Somalia or a biologist in Papua, New Guinea. In other words, a broad social learning process must be expected and must be counted as a part of social development before creative scientists can be expected to arise. Traditional and rural communities, beset by historically deprived learning environments and long-term lack of opportunities to grow intellectually, simply cannot suddenly leap into the new world with a crop of Nobel laureates. In this sense the Renaissance and Enlightenment may be considered as training exercises preparing our society and the varied classes of scientists, explorers, philosophers, inventers, and traders to arise and become appropriately assertive.

Indeed, through its own self-development, modern urban community must provide appropriate growth stimuli for the self-development and creative freedom of each person, as well as for itself. And this condition, more than any other, speaks both to the most critical role of a modern community and it contrasts most sharply to a traditional meaning of community. Thus the human potential becomes the goal of community as the chief instrument to achieve the system of *optimum social rights*.

Prospectively, therefore, community can become the quintessential force in a smooth process to develop new social rights as basic elements of human progress. A healthy community can demonstrate over time vast human advances, along with the family, being the heart of human development. The physical and social nature of modern community, especially when focused on self-development, makes it a fundamental educational institution. Each community therefore constitutes a basic internal diversity, really an extraordinary diversity close to people's daily living. And, being unified with a progressive outlook, the community's integrative abilities offer a basis to be articulate and flexible in initiating the best conditions for human growth. Thus, in both structure and method community can become an exceptional tool for human development and the centerpiece of a new realm of social progress.

Now, following upon the exercise of social *goals*, we can in due course build them into human *rights*. That prospect was set forth in the Bill of Rights. With amazing prescience, they also wrote in Amendment IX, "The enumeration in the Constitution of certain rights shall not be construed to deny or disparage others retained by the people." In other words, human development and the growth of social rights is embedded in the original concept of rights. New laws having self-evident values as rights and adopted in legislation at any level of government down to and including community, carry presumptive constitutional validity as rights "retained by the people." The Ninth Amendment is not only open ended; it establishes in effect a frontier for the growth of social rights. Consequently, goals can grow and become rights in a diverse national setting as a natural democratic evolution without excruciating debates and social disruptions that seeking a new Constitutional amendment from scratch would normally entail. The rights are already there, waiting to be defined and activated. The Framers have thus challenged us to grow. At this juncture of history, we can hardly do less than profoundly reconsider the content of progress.

Removing Blockages

In the many historic fluctuations of human worth, the record of massacres, brutal suppressions, and wars have often given society as a whole a gross cynicism about the value of human life. American democracy was born with slavery and this depravity was buried in the law of our democracy. Slowly, democracy has broadened the franchise and the rights of minorities: blacks, women, other ethnic minorities, children, and the elderly. These steps have removed a number of blockages to the growth of freedom and the potential of self-development.

Yet many serious blocks remain, and only limited progress can be expected until we can generate new visions of the human being in settings of expansive freedom. First we must note how Freudian psychology seemed to give proof to original sin. That belief remains powerful in degrading a vision of human prospects, but has been challenged by the growth of humanistic psychology.

However, greater blockages remain deep within the social structure. Conformities of class through inequitable wealth mean that social class is largely wealth defined. All persons in all classes are diminished by the many closed compartments and the massive, costly class defense systems that consume so much personal energy, and force society to focus upon wealth-defined classes as a central legitimate social value.

A corollary impediment, the contradictory "privatization of social life," is really a pay-for-participation system, and it creates a structure of suburban loneliness and a denial of humankind's social nature. Over the decades public social spaces have declined, especially in the decline of central city, resulting therefore in a creeping privatization of social behavior. Today "public" spaces of the city consist largely of commercial strips and malls that lock up at closing time and exclude all activity except buying. Similarly, many seeming social spaces like restaurants, hotels, and motion picture theaters, themselves highly isolating environments, are also devoted solely to paying customers. True public spaces are scarce in our affluent society. It is amazing to see how much of the city's spaces are obligated to driving and parking of automobiles; they thus guarantee a "perfected privacy." When considering public spaces, they consist primarily of huge paved spaces for parking, streets, boulevards, and freeways, of which the public, rather than creating environments for varied social participation, is co-opted into promoting the mobil portion of the perfected privacy.

The social life we do have is specialized, periodic, dispersed to the metropolitan reaches, and requires exceptional motivation. Consequently most people hardly belong socially at all, a classic condition of our urban areas. Churches are devoted to specific congregations mostly on Sundays. There are halls and convention centers and these, too, are special to particular groups at special times. Schools have perhaps the largest variety of public spaces, but these serve no more than children that are highly segregated by age and subjected to the preparation for a work career. Non-school uses are confined to rare events or perhaps recreational basketball. Local parks are minimal, most regularly used as tot lots. Some cities have active neighborhood centers; these serve valuable services of a limited kind, usually from converted houses, churches, storefronts, or schools.

Defenders of privatization of urban spaces can claim that there is no demand for more open spaces and public facilities. This merely reflects the fact that there is no demonstrated *commercial* market for them, and such interest as exists is focused on a promotion of single, specialized facilities like large classic museums and zoos. This condition illustrates the urban disarray and the fact that campaigns for new public facilities requires the struggle to create a huge public movement to overcome the normal commercial opposition to public bonds, especially when they must overcome the exponentially difficult requirements for a two-thirds voting majority. The virtual absence of a strong social presence merely reflects a deprived public condition without progressive visions for a greater social potential.

These conditions underscore the dismal fact that in today's society there is no public ethic for varied, highly participative facilities and programs, except some arts and crafts classes. And, if there was such a popular demand, it must face up to the city's universal miscellaneous dispersal of development in which each facility built becomes a single-use in a dispersed and isolated setting for activities without an organic connection to complementary activities and services. That is, there is no socially attractive organic whole that is easy to access in which an interwoven urbanity can result—qualities that a community could achieve with its integrative design. So we live in a socially depraved condition in which liquor bars become the demonstrated market as social clubs of last resort.

Altogether the space allocations and the larger questions of mal-formed cities also reflect the subordination of the public sector. The commercial domination of massed human economic particles is the preferred way of processing and marketing all elements of economic importance, including people as both employees and consumers. This means an underlying chaos of human affairs. What appears normal today, that is, the economic fragmentation of all basic elements of living, is in truth an indifferent, abrasive, and rude subordination of human affairs to dollar exchange. Human beings, since there is a long coercive tradition of human submission to cruel indignities, can never triumph when they are reduced to random commercial elements of a socially disorganized life, whatever the compensatory bribes. And so the behavior of human social interests becomes submerged and scattered in disarray, a condition that people have had to accustom themselves, as it suffocates their humanity and reduces them to imperatives that must be tolerated. Thus, when closed out physically and emotionally, people do not easily aspire beyond production and consumption.

The meaning is that humankind must find a means to extract itself from economic processes. A new, moderated work career might be appropriate. A second, non-economic career may be considered. A guaranteed income is essential. Social goals point to a life freed from the arbitrary forces of economic determination. Indeed, goals developed as social rights stress this possibility. Thus an economic order with reduced human burdens will also be more socially efficient and fruitful.

If we are to give preeminence to the social contents of life, such measures will require not only a significant withdrawal of society from its economy but also to a subordination of economics itself, as we have already done with agriculture. The behavioral processes of society belong to the people, not corporations. We need to think seriously about installing the public interest and its major instrument, community, as central players in such a social transformation. That means reducing the present powers of economics and directing clearer support to the true sovereign in society, the person in vital association.

As it is, there is relatively little basis for self-development because nearly all forces of urban and social organization serve, or are subordinated to, economic purposes. Psychologically, a person now tends to see and feel what one is expected to see and feel. That is because economic determination organizes social affairs and makes them seem like a law of nature. Since one's inward self reflects very articulately one's outward experiences, this scours out other states of mind, especially the freer social and personal initiatives of life.

Avoiding Social Rigidities

In 1789 at the beginning of the French Revolution, Jefferson in Paris had discussions with Lafayette concerning the rights of man. In the struggle to define what human rights should be, many ideas were considered. And in this span of ideas was, as Jefferson put in a letter to James Madison, "Whether one generation of men has a right to bind another." He said that, "by the law of nations, one generation is to another as one independent nation is to another." Each generation of about nineteen years could claim that "the earth belongs always to the living generations." Consequently all debts, laws, and even constitutions should then expire, leaving the new generation to build its institutions fresh. (pp. 110-111 *The American Sphinx,* Joseph Ellis)

Madison wrote back about Jefferson's "interesting reflection," then proceeded to cast them away them away as an impossible fantasy, as indeed they were. The point was that at the time of two great revolutions Jefferson was casting about for all possible foundations to liberty. In practice, of course, there could be no practical way to determine a precise generation and no way that fundamental institutions could be rethought through and rebuilt, especially by young, inexperienced youth, every twenty years or so. Yet the point about what is a beneficial heritage and what is dead overburden for each generation is at least a most worthy question.

This kind of thinking also reflected upon Rousseau's concept of freedom through the general will and the individual will he had set forth in *The Social Contract* some decades earlier. So, the question remains, can a new generation foresee its liberty sufficiently to strike forth anew? Can you become mature to see the world as it has never been seen before? Can society find an institutional means to avoid locking and confining each new generation into a social straitjacket?

Certainly, each generation cannot disinherit true human advancement, including a growth of freedom. One cannot easily distinguish what social practices are desirable to maintain, what should be replaced, and then create what should replace them. Nor is it simple always to know with certainty what is enabling from what is limiting, especially as some certitudes change. How can each generation, as its central question move confidently toward what are the greater human possibilities?

Today we cannot avoid asking, despite the experience of two centuries, whether democracy is practiced in its most generous terms. And given the growth of power of the corporations, which in themselves are not democratic, they can and do control the general course of public action in their self-interest. Therefore, it is impossible to establish reasonably valid terms of equality and justice anew each generation, even in the simpler terms of the eighteenth century. We all live upon practices established over the years, especially since our minds are incapable of changing more than a minute range of habits each day. The same condition is even truer of a whole society. The aim, therefore, cannot be arbitrary destruction and renewal each generation. If our aim is greater freedom, we must build foundations upon which new and greater freedoms can arise. The potential is to "give" every person a fresh view of life in which new personal growth is always possible.

However, Jefferson's question cannot be dismissed. If we value it as a search, I believe we can do no better than to translate it into the terms of pursuing a regularized system of goalmaking and a continuous public dialogue leading toward a concept of progressive social development. If our age is one of immense social change built by steady invention-driven initiatives, our chief problem has become one of giving all change agents of society a human direction based upon the highest social perception of the good life. Goals based on broad-ranging debate are essential. Furthermore, they offer a remarkably positive answer to Jefferson's concern.

If goals are sought assiduously, they offer each new generation the opportunity to retain what is best, adjust what is good, renew what is malformed, and replace what is destructive or dangerous. An effective goalmaking process can also protect the best foundations of both democracy and capitalism. Some people may argue that such a dialogue exists today in our legislative processes. That is true in theory but is seriously compromised by power politics. It is not satisfactory on at least two counts: First, it is overwhelmed by autonomous economic forces operating at all levels of society, especially from powerful influences over the consumer, power operating at elections, and special interests affecting legislation. Second, traditions are built around governance and social order rather than setting the course of change. At best they operate to resolve a few of the most visibly damaging issues that arise from change.

Goals, by contrast, are future oriented, change directed, and founded upon visions and ideals of what society might become. They are best suited to function freely above specific legislative action, striving to achieve general understandings and sort out public values before serious legislative work is undertaken. As valid as our Constitution basically remains, it was never conceived to be a constitution of change, and this is reflected in the serious weakness of legislative performance in our time of massive-non directed forces of transformation.

Yet, in one major arena, at least in the Ninth Amendment of the Bill of Rights, the Constitution does provide, as noted, a means to give direction to change—precisely in the growth of human rights, as I propose, through goalmaking. We can take heart, when it says that "certain rights, shall not be construed to deny or disparage others...." That is a sound foundation at least to begin a major process of goalmaking, which itself is likely to be the closest thing we can find to a *constitution of change*. That range may include prospective social rights described early in this chapter, most of which are closely allied to the development of community.

While the central thrust of the Bill of rights was a restraint of arbitrary governmental power over the free exercise of liberty, the rights now coming into view require, first, a reordering of those parts of existing society that now hinders a creative evolution of society, and, second, the construction of new rights directed to a new content of progress. They are a positive, creative, social, economic, ecological and open-ended growth of rights that depend upon public-interest initiatives as their basis for development. They require being organized into the normal processes of society to form a "nesting" place if not "a true re-creation of society" each new generation.

Social Renewal

Today human freedom rests upon two formal foundations and now reveals the need for a third. Upon the original foundation, the democratic process, another has been added, a material abundance, which relieved human beings from the huge necessity imposed by nature. However, relief from the necessity of nature has been replaced by another necessity seemingly as great, that of economics.

That substitution has occurred with an unprecedented intrusion of economic determination throughout social life. The tradeoff, as fruitful as it has been, is accompanied by domination of social life by economics that would likely strike Benjamin Franklin as a badly misplaced indulgence of the philosophy of Poor Richard who sought prudence and conservation along with the prime virtue of good work. In place of that thrift we now promote an incessant growth of producing and consuming, what Galbraith described as running in an ever faster squirrel wheel.

That domination of society by economics may be history's first invention of methods to socially subvert people to a single universal *system* of behavior. But that system, perfected in the twentieth century, combines not only technology and economics, but also controls entertainment, sports, news, and new methods of social persuasion, all bound together by vast expert bureaucracies that nearly succeed in becoming all things to all people. And it does so with surprising effectiveness, based on a doctrine of money motives and a comprehensive way of life that few are able to refuse. To loosely recall Rousseau, we fling garlands of flowers over the chains that suppress us.

A relief from economic necessity might one day inspire us no less than either democracy or the rise of economic abundance once did. And, if that inspiration one day might revolve largely around a concept of community, we might foresee new freedoms as well. If communities are established as small, face-to-face, self-governing, self-reliant societies of voluntary membership, they may carry with them a new source of specifically *human* development. Doubtless community will do best when governed by a process of self-development, for itself by and for its members. Each person and each community, therefore, might become a compact for mutual growth; one is incomplete without the other, and the strength of one contributes to the strength of both. This might constitute stage one of a process of growth.

Community self-development is founded upon conditions that create, integrate, and empower itself and its members, although certainly *not* the will of the person to be bound to the will of community, as some philosophers so wrongfully proposed, which led to totalitarianism. Rather, the will of community and the will of the person are mutually supportive, although community must sustain as broad a range of personal freedom as possible. New internal organizations will appear and request that their new community support them. Thus the community can begin to form new goals to serve its members, setting those goals into a priority, some as one-time projects, like installing a bronze statue in an alcove, and some continuing programs like organizing recreation for toddlers.

A second stage of community self-development, after a decade or so, will see a mature diversity in community, though still evolving, the appearance of the first generation of youth who have grown up in the process, a track record of volunteering and leadership, a first major perspective of how the community has progressed and where it wants to go, and a an common recognition and pride of local traditions. The community's vigor and accomplishments can be assessed with clear perspective.

Naturally, individual persons are the overriding objective. They also provide the most important feedback, first as to their own development and, second, to the creative contributions they make to the community. The community is small enough for the community to review its roles with individual persons. Behavioral problems can be considered in relation to the opportunities offered.

Nurturing New Freedom

Self-development means that, with all due assistance persons will have, the person takes the major responsibility for her or his own growth, beginning with making choices from the youngest age, becoming aware of self-learning through the years of youth, and taking full charge of vocation and avocation careers as full maturity shapes up. The community's role with each person is a continuous process of stimulation and support bolstered by assistance and many opportunities it offers. Self-development of persons is the central means to expand the course of human experience through the range of new and growing personal freedoms. Thus community becomes democracy's chief instrument developing the immense but yet largely unexplored range of human possibilities.

Already we are today accustomed to making a growing span of choices and career decisions, unlike the unchanging character of rural life until two or three centuries ago. We now determine what career role we want to play in society. And in doing so we also adopt personality-defining involvements by our choices of careers, churches, organizational memberships, and special-interest groups.

Nevertheless, even as we define ourselves by our major choices in life, we still define our personalities as largely rigid outcomes from those choices. We have much more freedom today than our forbearers in gender-defined roles, marriage, careers, clothing, and other behaviors associated with current affluence. Yet, while we are freer in personality terms than our predecessors of 1800, we remain less free than we would be if we had pursued a course of self-development within a freer social environment specifically formed to promote a freedom of personality

As it is, however, only a relatively few people are psychologically capable of beneficially engaging in significantly freer living. Similarly, since our habits are built so rigidly around money and material goods, those same social organizations—schools, corporations, churches—are unable to found a system underwriting a greater behavioral freedom. Society as a whole is not experienced in offering great freedom outside the realms of technology or economics. Each of the organizations with which we are associated largely defines the kind and level of freedom we enjoy—and by the same terms set the limits to freedom. Over all, how far we have come in expanding our personal horizons beyond those of our grandparents offers a clue as to how much farther we might go.

A large part of the growth of human freedom embedded in our psyche arises with cultural initiatives of new organizations like those of the human potential movement. So, if a new scope of freedom is to grow and offer greater depth of psychic vitality, an established self-development process with a new span of opportunities available in community could offer a basis for the entire human vision to expand. If past growth of freedom centered mostly upon technology and economics, a new freedom can be founded, I believe, in new forms of association and a new willingness of each generation to explore, evaluate, and add to their repertoire of human freedom. In community we can seed and cultivate the new freedoms as a common enterprise of all people of all ages. This, then, is our response to Jefferson.

Now we are accustomed to making major career choices and changes, mostly after high school. But since our education is confined so rigidly to career choices in both subject and uniformity of instruction, the tradition of the schools now confines rather than expands the human outlook, aside from accumulated knowledge. We are trained to be a "good citizen" but in such a narrow framework we are blessed only with methods of getting on as citizens in society. Learning to broaden our outlook is opposed by the current vision of public education. Personal freedom and initiative is still only rarely heard of in classes throughout the primary and secondary education. One can go out for a school sport or audition for a school play, but little more than selecting a topic for a term paper is available to students today.

If rigidities of the past existed through illiteracy and the confining ignorance in small, rural communities, we can argue tellingly that the confinements presented in today's schools presents a severe restriction upon human thinking and behavior so that we can all conform to the productive and consumptive models of human work, success, and demonstrative wealth. In the past during the race to production, that model was virtually the uniform goal of every person. It gave the narrow focus of the schools their legitimacy, their rationale to target the 3Rs in the lower grades, and skill development or college preparation in the higher grades, and to define them as the major achievements of promising youth. The goals, and therefore the freedom they expressed, were narrowly defined, freeing the person for an optimum career, but sharpening only those subjects and skills that promoted both individual and collective economic advancement.

The case for a larger range of learning interests that can be embodied in the community—which, after all, is a richly varied instrument of learning throughout life—is that the community offers an unrestricted range for life-long learning, an immense diversity in opportunities, and is highly responsive to individual persons. Its greatest expression is in the worth of the person. Its discipline is mostly internal within the person, and so personal growth, not external discipline, is the objective.

Since the capacity to grow and to exercise a larger range of freedoms society might look forward to, I believe there might be three approaches to help overcome today's confinements and underwrite a new greater personal freedom.

1. *The Person.* Community may emphasize personal growth as inherent in the prospective goal rights proposed early in this chapter.

2. *The Community.* Since society is not accustomed to "offering freedoms" to its citizens and remains limited to protecting the political freedoms in our tradition, the community's own freedom to grow might be exercised with a conscious vigor, especially by providing a great range of learning and social opportunities for its members.

3. *An Air of Freedom.* Yet a new freedom is not likely to go very far until an underlying character of openness, social invention, and the many cultural facets of freedom are expressed in a community's learning programs, penetrating dialogue, literature, and public expectation, far more, I believe, than today's awareness and support for the Bill of Rights. The air and spirit of freedom must give legitimacy to both the concrete programs of expanding opportunities and the individual person's willingness to step forward, think anew, and strike out forthrightly. In other words, progress is dependent upon a conscious willingness to advance the social dynamics of progress.

The compact of a community and its members may be understood as a special, conscious, and forthright social contract affecting persons, perhaps fulfilling some of the hopes the Enlightenment had for the growth of education. Its purposes and potential can unite us in a new kind of progress. Humanity has a long way to go, and communities could be considered as conscious if informal processes for a new character of person in society to emerge.

Evolving Human Truth

Any reasonable level of self-development by a person will mean the growth of a permanent agenda of thought, the excitements of fuller experience, and an independent course of development that is unique to that person, the maturation of a personal philosophy for living, a pattern of positive social involvements, and an inherent level of inner and outer responsibility. Here are elements of a possible ideal condition surrounding the person, resulting from community.

These transformations might be considered as steps to higher levels of human truth, and may prompt a massive shift of society to build a new dynamic dimension of human value. We can perceive that such a value will offer persons a better control over their lives—both an inner and outer empowerment—resulting in more independent thought. In the greater individuality urged by J. S. Mill, we may expect a growing ability to exercise freedom, a greater diversity of creativity and, indeed, a stimulus for genius to prosper. Somehow, through today's higher education an unfortunate discipline of school and the defeat of stimulating thought, a powerful anti-intellect infuses us precisely when the mind should blossom most brilliantly.

In social terms, we can look forward to a richer interpersonal life based on the greater diversity of personality and creativity. Liberty can be expressed with greater vitality. This will mean a more plural society, a true diversity with a renewal of distinct cultural and ethnic values. We can then rejoice in the local, human scale of physical and social environments. The democratic process can be filled with opportunities and express richer meanings within and around every person.

Early on in life a child will feel, perceive, and know that self-development can follow its own course to learn and design its life, to consciously pursue each stage of discovery from numerous alternative paths. And society can know, as well, that self-developing persons will make better citizens, will contribute more to society, and on average will do so with greater creativity and humanity. The person's self-obligation is to take responsibility for choices it makes and to build personal achievements. Obligations to society are to build some of those achievements for one's fellows, for community, for society, and for civilization. The possible range of service to society is as great as the diversity of persons and their works, and may be accomplished without any particular attention to social "contribution," such as protecting a waterfowl habitat, or a plea for some human value not yet appreciated by society. Possibly some persons' most important contribution might just be themselves, living well, being role models of thoughtfulness, being perhaps passive but positive and sometimes inspiring.

Community's role is built around the provision of personal and group opportunities with stimulation and assistance to act upon their own vision of what is good and right for them as free and unique persons. The community will know its success or failure intuitively, for rarely does there need to be a formal evaluation. The person will know that the new freedom does not involve merely cheap, numerical quantifications of experience, raw ego exertions, peer-group animations, or demonstrative rebellions. To be sure, self-development represents society's faith in independent human intelligence and good will, and a justified faith if persons know, unlike so many people today who cannot say that society is on their side of life.

Their truth in life is their own. Peoples' inner truth reflects their personal dynamics of mind and is not beholden to scientific logic or public rationality. If some of that dynamic is conveyed to others, all are richer from that sharing, for it can be like a personal poem of beauty that all persons may find insight into others beneficial for themselves.

Prompting a maturity of self-honesty and self-responsibility, along with self-development from the earliest age is far, far more important than merely imparting knowledge, especially in this age of infinite information. *Induced learning,* rather than *imposed instruction,* inherently responds to personality and to the unique potential of genius in everyone; it thus represents a higher human freedom in every person. Enthused learning and the contagious thrills of the mind are close to being the greatest pleasures of living, and undoubtedly penetrate into all other human joys, for joys of the mind are inevitably a part of all other joys, for all experience is processed or associated intimately within the mind. In time, self-learning might overcome the insidious anti-intellectualism present today is that philosophy is demoted to but one more specialization.

Possibly a true magnificence of personality—perhaps a new span of democratic personality—can arise and to underwrite a fundamental human liberation. Self-learning, the independent mind, the unique personality, and the elevation of personal experience forms a cluster of the human possibility that points to an infinite range of existence. Since humankind is also the only self-defining creature and continuously exhibits new possibilities shaping a massive new essence of existence, the pursuit of higher and more penetrating experience is plausible and possibly imminent.

We might foresee a new human dynamic of both emotion and rationality in which each serves the other, striking a balance as an ideal in society. If that happens, perhaps then society might find expressions of an unprecedented freedom, a creative inner spirit and a fresh behavioral vivacity.

The foundations for these elevations of human truth are (a) an urban society capable of varied forms of organization, (b) a productive system permitting civilized leisure, (c) an expansive democratic order, and (d) an evolving system with growing visions of the human possibility. When these four conditions are in place, a new *human renaissance* will be underway.

Next Liberation

The implications of community, and especially those of self-development, carry enormous human opportunities. They invoke life itself as an inherent process of human rights.

We now enjoy to a reasonable extent most clauses of the Bill of Rights, and they protect us *from* a number of oppressions. These are our negative assurances for a good society; they offer no direct positive benefits beyond a measure of security. Moreover, they are political and legal, not economic or social or cultural.

Our time presents us with a vast second challenge: the creation of the range of positive freedom *for* higher human opportunities and achievements. Yet wealth is the means for a wider, open, and enriched living, not an objective in its own right, as it has been since the time of general scarcity. Community is well suited to offer numerous positive opportunities that the new rights may be founded upon.

And another even greater range of human freedom is implicit in personal self-development. This third overarching freedom, based upon the foundations of political freedom and economic abundance, is *of* and *by* personal freedom. It reaches into the essence of choice formulation itself, one might say into the very "creation of human will." This freedom cultivates the very motivations underlying the greater freedom and exists at the genetic core of personality.

Thus, the paradox of self-development reaches its highest fulfillment close to the very heart of all that is human, our striving to become something other than that in which we find ourselves. This new geography of freedom presents a fresh, still unexplored scope of *what we are*, to more of *what we can be* as human beings. Inherently as human beings, we are very dissatisfied unless we can be striving for new reaches in life. Every stage of our development therefore requires new visions, new organizations, and new ways of looking at life itself. The discoveries we seek are of necessity largely within ourselves, for that is where all of our potential exists. With trepidation, I call it the *ultimate* human liberation, even as there must be more. Yet that vision also speaks to the paradox of what human beings have sought since they first left the caves to construct their own dwellings.

But such a liberation will remain imprisoned within us as long as society pursues its monolithic goal of what is the truly dangerous human paradox, the direct contradictions now explicit in unlimited and increasingly barren economic growth pursued only for its power. While the existing level of output can easily free us from the economic necessity, that growth overwhelmingly dominates the social mind of the highest leadership in government and business. In dominating our mind, simply indulgent growth harshly confines us to the great machine of productivity, promoting the endless reach for more, a vacuous nonsense, but nevertheless an ominous tragedy in the making. What should free us imprisons us.

A vision of a fundamental liberation is our first, best entree to a more direct human form of achievement while on earth. Community, I argue, offers a means to propel and unify, perhaps not an endless crusade, but certainly a grand challenge to the human ethos of our age. And self-development offers a means to embed progress directly within the person.

New Class of Freedom

Community, we will learn, is not merely a friendly association, for if properly developed it can offer society an entirely new range of human freedoms. Potentially a new ethos of human progress is foreseeable, a progress that lies within every person and emanates from within personality.

Human beings have been indeed fortunate through evolution, inheriting a body and mind with the capacity for immense freedom. Among the inherited qualities are walking upright which freed the arms to lift and carry, and hands to manipulate the "things" that resulted in civilization. We also inherited a voice to articulate sounds and build meaningful communications. Most importantly, our brain enlarged, giving us a mind for dynamic thinking and the creative uses of our arms and hands.

Those features underwrote an entirely new class of evolution I describe as superevolution, that is, for cultural development and civilization. This was first founded upon the development of complete language and a standardized vocabulary to generate the complex and beneficial concepts. Families evolved that resulted in training for each new generation and the accumulation of knowledge. In time writing appeared that created the possibility of communication with people not present, either geographically from past or to future generations. A division of labor then appeared that led to greater skills and higher productivity, especially capitalizing upon the domestication of plants and animals and the uses of wind and water power.

These two ranges of human power and freedom, the evolutionary and the super evolutionary, are a prelude to two major modern reaches of human freedom: first, our constitutionally based democracy with numerous specific freedoms founded upon the Bill of Rights; second, the industrial power that frees human beings from physical toil and the meager peasant compensation. So today the political and economic benefits acquired over the last two centuries leaves mainly a psychological and social gap in empowered freedom.

The question arises, therefore, whether the potential of community can over time offer the foundations for a human face of development to complement and perhaps earlier achievements advancing society? I believe that possibility may be best foreseen in varied freedoms potentially arising with community. That human advancement is foreseen as *social* and *personal* in its nature, *positive* and *progressive* in its achievements, and *universal* to human interests.

Freedom in community arises from its being highly articulated to the human condition, as government is to the organization of society, as technology is to systems of engineering, as enterprise is to products and services. What has been so glaringly absent in human progress to date is a social system that protects and empowers the person *as a person,* not merely to possess new wealth and vast congeries of goods, but for the person to live dynamically with others while in control of her or his own being. It means that society is not only democratic and wealthy, but that *persons will become the greatest achievement of society,* one who rises to the level of virtu. Could it be that a virtu of the person takes sight upon an ultimate range of freedoms? Let me list some specific possibilities:

* A vast range of personal involvements in a wide variety of activities but minutes away from everyone's door.

* Close access to employment, commercial services, public facilities, and organized social programs.
* Close access to a wide range of open spaces and outdoor activities.
* Close access to casual interpersonal behaviors, participative association, and organized behaviors.
* Numerous levels of valid intimacy, familiarity, and specialized or professional involvements.
* Greatly reduced need for mechanical transportation and no necessity to own an automobile.
* Varied forms of beauty in the built and natural environments.
* Community as a universal educational institution, both natural and built, formal and informal, from the cradle to the grave.
* Community as a comprehensive health care institution, preventive and corrective, minutes from everyone's door.
* Directive involvement by everyone in the actions of the community corporation.
* A combination of all freedoms sought by each person.

These specific freedoms offer foundations upon which greater freedoms can arise over time. Greater freedom can be derived from many new efficiencies afforded from compact spaciousness at the community's human scale, compact development that paradoxically increases spaciousness in both the built and natural environments. Similarly, an integration of many activities (as opposed to current practice of segregation and separation of functions) can become functional efficiencies of both individuals and institutions. Efficiency is basic to freedom because activities have ready-made facilities close together, avoiding distances, driving, time, and money that otherwise must be subtracted from each desired activity, be it theater, botanical garden, or genealogy.

Personal and emotional security also arises as a new freedom because a person can then interact more freely with others. These conditions in effect emphasize the person and personality while avoiding the impersonal involvements and alienation characterized by so many of today's organized activities. The places, the people, and the activities easily become familiar, even to a child. That continuing familiarity helpfully bonds peoples in mutual support and security.

The greater personal freedom grows with the unprecedented variety of personal opportunities for all persons to select and build upon for themselves. This dimension of freedom is expanded when it serves the person's self-development throughout life—the system expanding one's meaningful experiences designed by and available to all persons. Although the person is the focus of development, it is through close, convenient, and non-threatening involvements in community affairs in which the person can develop clear, self-confident, diverse, and spirited strength of character.

What can evolve over the coming decades is a true human determinism, taking leave of the economic determinism now so powerful among us, which in turn took leave of the determinism of nature in modern history. A human determinism can free humankind from what have become arbitrary burdens that resulted with the great wastes of the runaway economy. That transformation can help generate a true renaissance, possibly the greatest range of civilization ever, in which all persons may share all possible human achievements.

And if people as a whole are the focus of civilized achievement, they then have the possibility of attaining virtu, the highest personal achievement of civil society. And if community becomes a key player in that achievement, it too might be eligible to become acropolian. If virtu is expanded from a love of fine objects of art, to the art of being a superb person, the community itself might similarly become acropolian for its attainments leading toward the essence of the human possibility.

Thus entirely new classes of human freedoms are brought into social perspective, comparable to and complementing the political freedoms achieved in history. Since there can be no greater idea than being human, it is the repository of all social value. What community can do is directly generate freedom into the very being of every person.

Fundamental Solutions

The aim of community is to elevate human life as purposefully as we now build products. However, elevating life must serve human diversity and self-development. But, unlike product output, the benefits of community become catalytic and socially enriching over time. Physical benefits become mutually reinforcing with increased social values; social benefits arise with improved environments. Conversely, today's crises are social consequences of two centuries of single-minded technical and economic development in which the multiple social and ecological problems accumulated are in effect promoted with exploitive economics and consumptive urbanization, followed by escape to the suburbs, which accelerated the exploitation and overburden. That single-minded method of pursuing single-purpose economics reflects the inability of modern society to pursue comprehensive, holistic, and inherently diverse human interests.

The dynamics of community are found in its mutually reinforcing benefits. The human scale, for example, is both socially propitious and ecologically valuable. The combined results are therefore harmonious, unlike the multiple decadences of currently promoted consumption, waste, and unending functionalism.

Community is a system in which multiple human values derive from a continuous and growing articulation of spaces and facilities for all of its members, yet operates through a unified approach that simultaneously serves multiple physical, ecological, and social purposes while incorporating technical and economic methods and values by their least-means modes. Progress, therefore, no longer needs to operate upon unlimited economic growth since comprehensive values, holistic methods, and broad human outlook supersede the value of raw promotion of production and consumption. Here are some illustrations contrasting the potential and present conditions.

1. The community's concentration of created spaciousness with many varied benefits against today's low-density scattering which makes multiple family automobiles mandatory while also and resulting in social isolation.

2. The community's conservation of land, permitting unprecedented and varied open spaces close by stands against the present systematic wastefulness of land with minimal accessibility and limited variety of cultural institutions, especially in sprawled suburbia.

3. The community's direct and immediate access to a full range of spaces, facilities, and opportunities are set against today's social deprivation of social behaviors that create an overwhelming TV-dependency.

4. The community's optimal pattern of transportation (auto, transit, and walking) contrasts against forced long-distance commuting by car in ugly and inhospitable environments.

5. The community's creation of a social dynamism through many services easily accessed while today's services are few, scattered, and meager.

6. Community conserves land, which minimizes the need to travel, while today's promotion of consumption is costly, cumbersome, and ecologically destructive.

7. Community conserves money, material, time, and human energy, while the present system isolates people, denies association, and creates conditions leading to economic suicide.

8. Community operates abundantly with minimal consumption of resources, but today's economy grows upon its own excesses of the runaway economy.

This list could grow much longer. The immense contrast between community and the contemporary system means that while social change now revolves around one purpose, that is, today's technology-driven economy, other purposes inevitably suffer. So, indeed, we suffer major and multiple crises of society and ecology. That is the case for a new kind of diversity focused upon the person which society can achieve.

If freedom is advanced by the community and by its very human building, it can also resolve the two general crises confronting society, the social and ecological, while opening major new positive human possibilities. Today's crises are created conditions arising out of the technical and economic conditions that are endemic to that long-term penetrating development.

If the aim of community is to elevate human life as purposefully as we now build products, we will then be able to address the social and ecological issues mostly in the course of pursuing higher human possibilities, since these crises are also embedded in the conditions we confront, to replace urban inefficiencies, the movement-dominated cities, the systematic waste of urban space, and the social generation of crime and bestial human destructiveness. Thus, as we seek more humane environments we can simultaneously resolve many other specific problems of life in cities.

While my primary aim is positive human living conditions, it was a pleasant surprise to discover that I was also simultaneously addressing the same dismal questions that afflict today's society. But they appeared, nevertheless, as outcomes to the positive pursuit, incidentally it seemed, although not as basic objectives. And so, if my perspective is valid, these crises can be resolved mostly in the course of pursuing positive social development through community.

The savings of rural land incorporated into cities can drop by 50% to 75%, depending on the scale of urban sprawl. Furthermore, up to 80% of the land assigned to community can be applied to urban open spaces, including golf and other recreation, parks and lakes, urban farms and forests, botanical gardens and ecology study areas.

Although anyone may own automobiles and use them as they will, there will be relatively little need for them, none within the community, which will merely provide parking in its basement. Motor fuel savings will therefore be enormous; the community building itself will reduce heating and cooling energy needs by 50% or more. The generation of pollution and greenhouse gases will be reduced proportionately.

The core of many of today's social problems is found in the disasters of urban forms and functions, compounded by the forces of escape and urban sprawl. Just one of the crises is that of desperate freeway congestion and the paranoid proposals for trillion dollar "relief" construction, but such construction will merely increase the congestion on the boulevards and parking facilities upon which the freeways depend, as it has always done, only make the anti-human city more inhuman, more costly, and absolutely unsustainable. *We cannot construct our way out of the auto dilemma,* as we have struggled to do in about five campaign cycles since the 1920s. It is irrational to promote traffic to another level of public horror at a horrific cost and further loss of human environments.

Our current urban form is inhuman and respects neither the physical nature of people nor the psychological and social nature of a healthy society. Schools are seriously defeated; ill health and accidents are generated, all while we destroy the foundations for a sound psychological health. Astonishingly, we have an advanced combat-ready ambulance and rescue service but fail to strike to the core of auto accidents, family violence, and unhealthy minds.

Economically defined social privilege, economic desperation of the poor, rigid, ill-founded education, broken families, teen-age pregnancies, and the consequent depressions of slum living are the gross conditions underlying drugs, violence, and insistent crime from generation to generation. The larger cause is that we have failed to seek a joyous social life among people, despite the increasing capacity to do so over the last century. That condition can be reversed in communities that are basically small cooperative societies in an environment designed to be socially and psychologically healthy and filled with opportunities.

Communities will have major impacts on local economics, especially to reduce the massive arbitrary consumption of land, and the machines to reach it, occupy it, tend it. Consumption as a whole will be beneficially reduced by the community's shift in emphasis from *having and owning* to *being and doing*, all while preserving ample resources for product choices. The community will become a consumers' union, an organization not only to eliminate unneeded products, not only to reduce material needs or share facilities and equipment, but also to convert the popular grass roots to tap roots giving organized strength to the consumers' side of the market place.

Moderating consumption and organizing consumers will also have the effect of significantly reducing the oppressions of high-pressure commercial promotions, completely in the case of urban sprawl, and significantly in our product-dominated psychology. The needless work-dominated system of education since the 1800s denies a human creativity that could well serve both a professional career and a personal avocation. Yet, it is in community that the current economic determinism and its massive momentum in the runaway economy can best be reduced or eliminated.

Building human capacities builds positive human freedom, what communities can be designed to do. Human potential lies within every person and the purpose of communities is to help every person to internalize and externalize their greater capacity to live. For all of the community's physical benefits, efficiencies, facilities, and programs, they are also foundations for every person to grow in her or his own terms, to pursue her or his own self-development.

If the twentieth century foresaw an amazing growth of general wealth, that wealth now needs to be transformed to give the social and psychological conditions of every person their chance to grow. *The person must be supreme and community cultivates that supremacy.* In community is the objective of fulfilling the

democratic and economic advances of our age. It can become the human fulfillment organization of unlimited advancement.

The basic problem of development in our age is that economic growth has
been promoted endlessly as the one final objective for an economic nirvana. In
the process other basic realms of human progress have been pushed aside. Cities
have been promoted to systematically yet recklessly force-fed consumption. That
has made them human disasters of sprawled isolation and central city deterioration with freeways build to bridge the distances between home, work, shopping,
and recreation, but always destructive to neighborhoods they cut through, yet
inadequate in their supposed high capacities. Governments, pulled into the
process, have accepted the endless burdens of construction in the cities and the
endless growth of the entire economy, and so they have become accomplices to
the crises society now faces. *This monolithic tragedy misconstrues economic wealth,
destroys cities, consumes nature, oppresses human behavior, and results in the modern
system of social horror.* In the present terms of public action, there is no recognizable solution or prospective relief.

Community creates a human power for human development that can redirect
social behavior to solve the greater ecological, social, and economic issues while
offering new richly diverse opportunities to pursue a specifically human agenda.
It is time in history for humanity to direct its powers to a balanced, harmonious
process of human development. Humanity cannot improve if it stands deep in
the dismal effluent of wealth-made slough. Society cannot deny this tragic misdevelopment if it remains bound to monetary exchange as the ultimate good in
society.

Community, as I see it, is a society writ small—to the human scale where personality arises and develops its own naturally healthy structure—where greetings
are hugs, interaction is eye to eye, gestures are painted in smiles, and voices rise in
pleasure and end in a laugh. How often, where, and in what conditions is this
kind of behavior permitted and socially encouraged today?

Grand Harmony

I have pleaded herein the close unity of cities, democracy, and ecology. Yet I have
also argued the central instrumental role of cities, and emphasized the potential
of highly integrative communities for their essential human qualities and as
rational building blocks of cities. Communities could become the first human
institution to serve the complete range of normal conditions in the settings and
environments in which human behaviors are played out. While the companion
book, *American Communities,* in the present series considers the prospective form
and organization of communities from numerous specific perspectives; I have

limited discussion herein to their general roles and broad significance. So, as my case unfolds in both volumes, communities constitute a central concept of a new form of cities, emphasizing that cities are central to a prosperous democracy and a valid, rational role in both democracy and ecology.

Cities are fundamental. If society is an intentional creation, cities are the core of that creation. And, given the unfortunate, almost tragic conditions of cities, they must be basically reformed to permit a fundamental human liberation to arise. No longer can we leave cities to a thousand random occurrences in shaping their structure, character, and human impacts. No longer can we fool ourselves that mere *planning regulations* will ever achieve a valid *urban design* in pursuit of human goals in cities. Cities must be created as one of the most precious formations of our emerging civilization.

Why haven't a dozen ideologies vied for the conceptual designs of urban form, a thousand urban images of good living been conceived, or a central theme of literature raised the visions of urban life? Cities require theories of the social sciences, numerous philosophies to anchor human purposes in the physical and institutional environment, the creative forces of the humanities. Of course, cities require the rigors of ecology, the esteem of honored places, the locality for a superior human identity. But in today's cities these are for naught.

It is for us in our modern predicament of cities, ecology, and democracy to look toward a vast renewal of cities as the most basic sources for human renewal and a higher integrity. Democracy and ecology cannot advance far without a profound reconstruction of the human setting. Their clearest resolution of vast human *problems* is the achievement of a least-means economy, requiring a crusade to free us from the overwhelming economic determination we live under. The clearest human *prospects* will appear through consciously conceived community and especially its socially assisted self-development of persons. A new development of cities is close to the holistic completeness that modern life so urgently requires. Not to envision cities as central to human progress is to ignore the rational methods of science, the basic lessons of both technology and bureaucracy, the creation of universal efficiency, the accountancy of money, and the supportive role of economics.

Cities must be at the center of a new realm of human progress. They define the physical structure of public behavior, determine human efficiency, establish facilities and organize human opportunities. And, simultaneously, a renewed approach to cities will resolve the social and ecological disasters of today's human existence.

APPENDICES

A. *Language Of Cities*

Language is a tool of thought as much as communication. The vocabulary we use sets the direction, parameters, and dynamics of our thinking. Words reflect reality, at least one's sense of reality. As we know from conversation, words change meaning—and may sometimes lose their truthfulness. Some words may simply reflect error from the outset, as did bloodletting in the medical practice of the eighteenth century.

It is my contention that the city planning movement remains in a state of bloodletting at the end of its first century of its modern practice. This practice is reflected in the language of planning, which surprisingly well demonstrates the bloodletting that permeates it.

Conversely, the many years of work leading to this book has beckoned me toward a different set of words which I have used in search of the kind of city that I believe is worthy of our age. Many years were required to break with the old vocabulary, since I believed in it and tried to make it work. Reflecting back, it would have been impossible to refresh my language while continuing to work in the planning field, or for some years afterward. The sheer conformity demanded within a bureaucratic and professional environment is overwhelming. I discovered how my own thinking took on its own character and substance in direct proportion to the length of time since my last bureaucratic planning involvement.

Words can be beacons of light, or they can throw heavy shadow, confusion, and deception into thinking and communicating. Therefore, I believe, if cities are to be renewed or recreated to their core, both our thinking and the words we use as stepping-stones of thought also must be renewed. If we can say that we are what we eat, we can as well say that our thinking is our vocabulary.

New words with new directions and dynamics must establish a new reality in the making of cities. Words, like the concepts behind them, as well as their underlying methodology, require testing and evaluation, regularly and critically. We can retain those that pass the test of truth, as every era defines its own social truth.

The words below reflect new approaches and methods to be confronted and tested. Following my discussion of terms, I refer to some old words of city planning, suggesting why they bloodlet our social behavior, our environment, and our struggle for human worthiness.

Proposed Terms of Optimization

Each of the following five terms calls out some elements of the others.

Universal Efficiency. The combined optimal human effectiveness in production, government, public services, transportation, utilities, commerce, personal and group activities; thus saving time, work, resources, and reducing the need for product; especially to foreshorten the course to human aspirations; all arising from the structure of the built environment of each and all parts of the city.

Compact Spaciousness. Optimum unified, compact yet spacious indoor residences, commercial, and public facilities for a given population; applying the human scale with integrative design, which simultaneously conserves and therefore expands outdoor spaces and natural areas; decreases or eliminates the necessity of formal urban transportation (not by expanding it), private and public operating costs; and creates urbane socially interactive spaces.

Access. The combined index of urban transportation burdens of distance, time, travel mode, cost, energy, pollution, space consumption, and environmental burdens of a trip to desired destinations. Also considered are interests not pursued because of excessive access burdens.

Functional Integration. The articulation of varied functions and spaces close together so that all functions benefit by mutual proximity, and easy access for the people served.

Human Scale. Public indoor and outdoor environments and facilities, public and commercial services, and interpersonal associations inherently congenial for peoples size, behavior, and character to (a) be visually and personally attractive to people, (b) help achieve an optimum personal performance, and (c) promote congenial human interactions.

Proposed General Terms

Family Centers. A group of families, typically fifteen to forty, residing together as a living unit with common facilities and sharing such domestic activities as they choose.

Privacy and Publicity. Readily available options permit individuals to control the frequency, character, and extent of their personal seclusion and public interactions with others. Privacy without a full span of options and opportunities for varied public involvement is merely deprived social isolation. The whole continuum between privacy and publicity must be fully available if it is to be an urban right.

Urban Geometry. The creative discipline organizing the dynamics of three-dimensional urban space for greater efficiencies, inhabitable qualities, and social and ecological values.

Urban Algebra. The social organization of society suitable to enhance the qualities of living, including health, education, and other elements of personal development.

Least-Means Economics. Like universal efficiency, least means pursues general efficiency, but is restricted to productive output—classic economic efficiency.

Social Sciences of Cities. The active social sciences pursue applied experimental knowledge in building cities and institutions that serve public goals, bringing the social sciences into a parallel role with the applied physical sciences and technologies.

Urban Constitution. Statutes and adopted principles serving a new urban order in society, setting forth realizable goals, building legal and institutional framework for development, and organizing for the creation of new urban freedoms.

Public Goals. The constitution of change.

Goal-Oriented, Multi-Purpose Planning. The essential methods to create high-value cities that underscore durable ideals and expand the scope of democracy through positive, broad, long-term, and integrated programs of public action. This approach supplants today's monolithic goal of endlessly expanding economic growth and the crisis-driven, single-purpose, pragmatic, and specialized responses to problems generated in society.

Social Democracy. The expansion of traditional political democracy into the economic, physical, social, and cultural realms of urban life, emphasizing the expansion of human opportunities that build the positive new social freedoms.

Self-Development. The opportunities, encouragement, and assistance for individuals, especially youth, to pursue personal development according to their own developing aspirations.

Urban Ecology. The urban forms and functions that define broad social interactions between people and nature, always seeking to underwrite a harmony, permanence, social richness, and regeneration of nature.

Generalist. A discipline acting as a professional and scientific counterpoint to specialists. The generalist emphasizes human ends and social outcomes, the long-term and combined impacts of social action, the ethical and humane content of individual and institutional behavior, and the broadest range of knowledge and philosophic inquiry. The generalist is a philosophic ombudsman guarding human values and purposes.

Community. A socially coherent, self-limited population living in compact spacious building system pursuing richly varied opportunities for all persons. Community is the most basic urban unit and seeks to attain the highest levels of *social democracy, urban ecology, least-means economics, universal efficiency, and human esprit.* It does so by beneficially uniting and integrating the living environment, large and varied natural areas, public and commercial services, basic economic activities, a wide range of social organizations, a continuous serendipity of social interaction, local identity and pride, and a rich tradition.

Community Building. A unified, multi-purpose building facilitating the objectives of community, especially integrating the many functions and magnifying the varied richness of the human living environment, maximizing individual and urban efficiencies, and reducing overhead costs.

Community Space Frame. Same as the community building: (a) specifies community uses, (b) utilizes the dynamic third dimension, and (c) involves a large, articulate, and highly efficient structure. The space frame is intended to integrate many functions, to establish many human efficiencies, and to serve varied social purposes.

Community Corporation. The corporate instrument of direct democracy managing the community, providing services, unifying and expanding governmental, commercial, and associational roles of the person.

Urbanity. The public environment and social ambiance of community and metropolis that stimulates wide-ranging discourse, intense cultural vitality, continual interpersonal stimulation and growth, and an easy interplay of varied persons, organizations, and interests.

Runaway economics. Production and services based on commercially promoted needs not original to the people. These needs are created by building dysfunctional cities that require massive arbitrary consumption and consumption promoted by mass media. Solutions to problems generated by the runaway economy often add to the dysfunctions, demonstrating the power of economic determinism and the monolithic economic domination of social goals.

Current Urban Planning Terms

Zoning. The segregation of all major classes of urban activity. Zoning is intended to prevent mixed land use but in truth promotes urban *dis-integration,* that is, dispersed development and the need for excessive transportation. Zoning has validity on a regional scale, where it can define urban limits and protect rural areas from indiscriminate urban sprawl; but generally it has not been used for this purpose.

Subdivision. Controls conversion of land from rural to urban uses, principally through block and lot layout, with standards of street and utility development. Subdivision is a process of fragmenting large parcels into small, exclusive, individually owned parcels, serving market exchange rather than socially unified and integrated urban form.

Land Use. The activity to which land is put. The term emphasizes flat urban land, not dynamic three-dimensional urban space, and mandates a single plane to organize the city, and therefore induces an endless demand for transportation.

Population Density. The number of people residing per acre. There is a pervasive assumption that high density is equivalent to crowded tenement-style living or at least barren apartment-house occupancy. It reflects an historic paranoia, using the false analogy of crowded-rat experiments as proof. In reality, great urban spaciousness can result only from compactness of development, because the *high net density* and *low gross density* affords large natural open spaces, with close, immediately accessible services and activity spaces.

Housing. The residential building stock of a city. Housing implies high-density apartments, especially public housing, and therefore government providing lower quality dwellings as a welfare program for the poor. Thus housing carries a double pejorative of (a) public welfare and (b) high density.

Transportation. Urban transportation today mainly means automotive transportation. However, transit is making a slow comeback because of the desperation to avoid total auto gridlock. Walking is rarely considered, except as the means to reach one's car or transit station. Auto transport is sought as an endless good, even while its terms of achieving accessibility are abysmal. Thus the expansion of auto transportation is counterproductive, reducing access values while increasing social costs and family isolation, wasting the countryside and arbitrarily destroying resources.

Central Business District. The term, while denoting the urban center, reflects proprietary business interests, mainly commercial activities, ignoring other public and cultural activities, which give heart and pride to the urban core.

Congestion. The term normally refers to crowding through high-density housing. But modern congestion is almost entirely a congestion of movement and parking of automobiles, which results almost entirely from low density and the resulting necessity to travel long distances.

B. Taking Action For Cities

If democratic values of cities in these dynamic times are to survive and grow, only the people can speak to human aspirations. People must learn. They must find consensus and organize. They must act, and act fundamentally.

Cities today need popular action to secure and expand their human horizons. Change, of course, must always begin with the way things are. Yet the momentum of change must always put ultimate possibilities into focus and shift to the more fundamental goals when they become realistic. So ultimate solutions can come into view first by taking direct pragmatic actions.

People may work alone, writing, teaching, promoting. They may organize together to work on a dozen fronts of creative *urban* ecology. Today's conservation organizations need to develop sophisticated urban programs, if only to protect rural, mountain, and shoreline areas, along with endangered species and habitats.

Following are a number of lists of things to do, individually and collectively, now and in the long term.

NOW

1. *Learn/Teach*

   a. Stand on a boulevard traffic island for ten minutes. See and hear the environment you are in.

   b. From the roof of a tall building, observe the movement, the spaces given to movement and parking, and the resulting human quality of the outdoor environment.

   c. Attend a city or county planning commission meeting, noting the nature of the agenda, the discussion, the sense of purpose, the opposing arguments, the matters of urgency, the indications of power.

   d. Read the local general plan or special plans, question the planning staff about matters of interest to you, discuss what elements are most likely or least likely to be carried out, who carries them out, and how the plan is regulated.

   e. Visit the most progressive urban development project, also the poorest and wealthiest districts, and make as many comparisons and judgments as you can.

   f. Visit in and close to the old city center and the newest large shopping center and compare their functions, spatial organization, management, and the effects of the automobile.

   g. Find the best and worst walking environments in the city. Determine what makes them good or bad, and see if there is or can be any social content or meaningful personal value of walking along the way.

h. Investigate how public money shapes urban development, is forced by private urban development, and what the results may be.

i. Compare and evaluate public costs of complex traffic signals, freeways, freeway interchanges, parking garages, fire and police equipment and services, parks and recreation programs, school busing, etc.

j. Compile and estimate all costs of your automobile: purchases, accessories and repair, fuel, insurance, death and injury, driveway and garage, parking at work, other parking, streets and boulevards, freeways, time of commuting, and traffic delays.

k. Ask why (1) so many roads, (2) so many cars, (3) so much movement, (4) so much congestion.

l. Ask why (1) so much enforced privacy, (2) so much defensive behavior, (3) so much escape from central cities and (4) so much money for these conditions.

These learning experiences may be extended endlessly, made with greater depth, and organized into systematic research.

2. *Join/Organize*

a. A strong diverse urban-ecology movement.

b. Neighborhood associations

c. Land and nature conservancies

d. Cohousing organizations

e. Coalitions of museums, conservation, and other public-interest organizations

f. Urban historic preservation societies

g. Political action organizations

3. *Promote*

a. Beautification

b. Transit

c. Compact, integrated developments with large open areas

d. Classes on urban-development issues and possibilities

e. Urban design and sign controls

f. Critical land-preservation laws and programs

g. Strong, integrated downtown development

h. Social Impact Statements with Environmental Impact Statements

i. Urban development boundaries

4. *Oppose*

a. Scattered urban development

b. Urban freeway construction

c. Subsidies of scattered development: taxes, utility rates, schools, highways and other public services

d. Tax assessment of rural areas based on their urban development prospects

e. Zoned segregation of development

f. Height limitations

g. Parking-lot and garage requirements for new commercial projects

h. Subdivision sterility

i. Commercial strip commerce

j. Infringements on rural land, quality farmland, water sources, wildlife habitats, and natural beauty

LONG TERM

1. *Learn/Teach.*

a. The ideal urban potential

b. The organizational, legislative, and financial requirements for higher urban potential

c. Major urban design options

d. Promote study groups, produce pamphlets, and generate newspaper sections to publicize the urban potential, and establish a public readiness for basic urban reform

e. Promote school and college classes on the higher urban potential

2. *Join/Organize*

a. Societies whose aim is fundamental urban reform

b. Political action groups to promote the legislative, organizational, and financial foundations for the development of communities with compact, integrated, community buildings

c. To purchase or otherwise preserve lands for integrated communities

d. To promote urban plans and redevelopment projects for in-city development of communities by stages

e. To promote feasibility studies of (1) standardized, low-cost, industrial construction of community buildings; (2) integrative prototypes for community building design; (3) governmental, organizational, operational, and financial management of community; (4) social, cultural, educational, and health organization and behavior in community; and (5) organization, activities, and uses of community outdoor spaces

f. Seek federal and state support for experimental communities in various regions, topographies, and urban conditions

g. Promote community and the greater urban potential as the frontier national program for the new millennium, comparable to the lunar-landing program of the 1960s and 1970s

ALWAYS

a. Inspire urban science and ecology as a frontier knowledge of society, the greatest art of humanity, and the soul of civilization.

b. Compose music and paint for the higher worth of the city.

c. Beautify cities with public art, sculpture, mosaic, fountain, pavement design, and landscape.

d. Promote urban news, criticism, and visionary writing.

e. Celebrate, revere the best in cities.

f. Honor and exalt persons and organizations that improve cities.

g. Mourn, march, attack, and sue the worst in cities.

h. Excite children to dream, understand, plan, and work for imaginative urban environments.

SCHNEIDER DIALOGUES FOR THE HUMAN POSSIBILITY

Books have been historically central to human dialogue concerning the course of society's development. Whether the Platonic dialogues, John Locke's *Treatises of Government,* or J. S. Mill's *On Liberty,* books remain as timeless reminders of humankind's historic dialogues on the development of civilization. Today, however, with the great pace of change and the vast debris left in the wake of material progress, yet also the immense human promise afforded by that progress, the role of books and the vigor of broad social dialogue has fallen ironically to a low ebb in public affairs, narrowed mostly to technical, economic, or other specialized arenas.

The seven books noted in the front leaf are intended to jump-start a wider range of human dialogue focused directly on the human social condition, specifically upon the greater human potential. But unlike most social dialogues of the past that were based on a social elite, the current effort is directed to promote a far wider democratic participation. It also seeks to reverse the modern trend in which public discourse ironically has become seriously confined precisely in this era in which higher education, the supposed fountain of intellectual sophistication, has expanded to an unprecedented proportion of the population.

The seven books were not written as a series to systematically cover a given range of interest. While all were written to stimulate dialogue—which I define as conversations on important subjects—each was initiated independently to provoke thought on what I considered to be powerful issues that needed to be addressed by society. They present specific proposals that reflect my understanding of the issues and their resolution. The proposals are important, I believe, in advancing debate, in building concreteness into the discourse and, if nothing more, by stimulating alternative proposals.

One of our issues today is our specialized approach in discussing only pressing questions and responding to them with single, specialized, pragmatic solutions. Such a confined pragmatism limits public interest to resolving problems rather than pursuing positive objectives, leaving people in fragmented living conditions and society dangerously open to the powers that be.

Our era is the most dynamic in history, a dynamism that requires broad and clear social direction if the fundamental changes we confront are to fulfill the human promise, avoid human oppression, and avert ecological disaster. Our age therefore requires a new level of knowledge of social possibilities to establish an optimum course for society. Otherwise the forces operating today will all too easily undermine even our current level of human progress.

Today's dynamics of change therefore requires a new, broader human direction that spans the range of the social potential: cities, corporate economics, the form of human community, environments, resources, and most of all, an endless human potential.

In a democracy, dialogues are essential; together many matured perspectives can help establish a supreme but informal parliament of the people.

Through extensive dialogues we can build new inspired visions of life in society, visions going beyond the standard of living and its naked consumerism. At stake is the essence of human progress beyond production and consumption. Visions become a frontal quest for a fuller human experience of life for everyone, achievements previously reserved for philosophers, saints, and sages.

Such a potential already stands before us, based on existing technology, ample material wealth, and a growing freedom of time. The question uniquely challenging us is whether a new inspiration of life can be fashioned to reset our visions of progress upon the promise of our being. A high standard of living helps, but to expand the experience in life, such wealth is but a preliminary enabling condition. Nor has democracy given us beneficial directions for the greater possibilities before us. Therefore, a new method of public self-enlightenment is necessary to inspire the new, greater visions of human life on earth. The central possibility for such an awakening, therefore, arises in generating broad public dialogues about the human future.

The dialogues I propose will need to formulate many social goals leading over time to public adoption and implementation. Dialogues are thus a goal-seeking medium for higher human possibilities that go beyond simplistic dollar values and the democratic choices placed before us.

The questions of who, what, where, and how dialogues take place itself needs to evolve. I have made a few limited proposals. To be democratic and free, they need to be informal and arise from as many sources as possible: academic and professional, but mostly popular. As far as possible, they should center on ideas of development that reach to the essence of what human beings can make of themselves.

Through imaginative thought and progressive public communication and debate, the entire society becomes a forum for its own development. It might organize a new realm of the *public interest*, as I propose, to systematically shape social wealth for highly varied forms of human fruition. The public interest is potentially comparable to today's two great social organizing systems of capitalism and democracy, creating new environments and institutions with very conscious and specifically human aspirations. Neither capitalism nor democracy has revealed a clear orientation to pursue higher human possibilities. The public interest might therefore help overcome democratic society's weakest modern role, that is, to give positive stimulation and render creative direction for social development. Dialogues can become the catalyst for a public interest system of successive social visions and advances.

The forefront of the dialogues are not only the critical *issues* facing society. Far more importantly, they will consist of positive *ideas* to guide social development. Issues and ideas together can be understood best if they are conceived, compared, and acted upon for a common social interest. If the dialogues are to avoid the trap of today's practice of merely reacting to single pressing problems that are confronted with single-purpose, narrowly reactive pragmatic responses, society needs to set aside its self-limiting fixation on issues and its monolithic objective of endlessly expanding economic growth. Goalmaking through dialogues can shift society onto a new course of progress that is social and cultural, not merely technical and economic.

To penetrate society, goals that might displace today's single overwhelming objective of economic growth should be derived from defining clear human purposes through dialogue. Some of those include: developing new positive social rights; a personal liberation and self-development; reformed and socially exciting cities and environments; human community; and the immensely efficient community building. All are a part of the public interest as a broad human fulfillment system.

Of course, the dialogues must deal with today's issues as well. These can best be considered as a part of broader, socially positive ideas of development. Some of these issues are: economic determinism of life and the enormously wasteful runaway economy; the disastrous impact of the automobile on cities; the mass media that overwhelms the First Amendment rights; widespread economic oppression; the endless promotion of consumption; socially induced crime; and reforming education to serve the person rather than merely a work career.

Dialogues on the human possibility can promote a new dynamism of democracy. When widespread and highly communicated, the dialogues establish an informal, vigorous means to enlighten and give vigor to social development, assuring also that the content and process of change will be completely human, not merely technical or authoritarian as it is today. Dialogues can become inherently goal-formulating enterprises, articulating many human aspirations beyond the standard of living. They can influence change directly in human association, community, and metropolitan areas, as well as with the states and nation. Indeed, they can enlighten each person's self-development.

Human dialogues giving directions to development are vital when change is massive, powerful, complex, and rapid, which makes popular clear thinking and astute public guidance imperative. They can constitute a vigorous role beyond electoral citizenship in which people are active in conceiving, developing, and adopting the steps to a more generous society. Where in the past people worked for the economic wealth we have, people can now pursue a new, humane form of *social evolution*, no longer be restricted to its material limits. Thus human dialogues become the essence of a far more dynamic democracy, a leap beyond corporate political maneuvers and passive voting.

Whether a dialogue group includes two or two hundred persons, it requires discussion, learning, teaching, and networking. Its sessions can seek new knowledge and the wider perceptions of life now made possible, but which yet remain to be realized. The efforts of public discourse should lead toward insights essential to formulating important social goals, with a gradual consensus in formulating and adopting them, and perceiving the means to implement them. The promotion of dialogue will be deemed successful when it breaks out of organized sessions and becomes spontaneously pursued over lunch, in hallways, or driving to a public event.

There are numerous ways to gain perspectives and initiate *Dialogues For The Human Possibility*. The books noted herein can provide a start. Other books and articles also need to be brought into the dialogue as proposals, concepts, social criticism, or background. Other helpful materials are included in my web site: (www.KenRaySchneider.com)

Happily, a number of groups are forming to stimulate dialogue in small informal meetings. I note four:

Conversation Cafés promote conversation in public places "because when you put strangers, caffeine and ideas in the same room, brilliant things can happen." (www.conversationcafe.org—206 781 5700)

The World Café stimulates "conversation around questions that matter…sharing knowledge and creating possibilities." (www.theworldcafe.com)

Socrates Cafés and **Philosophers Clubs** sponsor "spontaneous yet rigorous dialogue…unique philosophic perspectives and worldview…for more deliberative and participatory democracy" by The Society for Philosophic Inquiry. (www.philosopher.org)

Café Utne promotes on-line conferencing with 80 open forums moderated by a group of volunteers. (www.utne.com/café)

Book Briefs: *Schneider Dialogues for the Human Possibility*

Book One

Autokind Vs. Mankind: An Analysis of Tyranny, A Proposal for Rebellion, A Plan For Reconstruction

The book stresses the devastating impact of the automobile upon everyone's life in the city, its occupation of much of the most critical human spaces, the vast reach of the automobile in promoting urban sprawl, the rasping of traffic on most parts of the city, and the abandonment of large older areas. The gross inefficiencies and resulting isolation of people contradicts the supposed utility of cars, reducing rather than expanding access—the traditional purpose and historic efficiencies of cities, that is, the fundamental nature of what cities are and must be. The book outlines a process of urban reconstruction based upon an urban geometry at the human scale and the conscious pursuit of higher human purposes in cities. These restore and improve urban efficiency, and establish a foundation for a grand vision of the urban potential to appear.

Book Two

Destiny of Change: How Relevant Is Man In The Age of Development?

Reviewing the course of social development in the twentieth century, *Destiny of Change* promotes new ways of perceiving the person, economic output, social community, the city, public interest organization, education, social goals, and democracy. Stressed are the pressures of modern living that now diminish human integrity, promote mass human subversion, and diminish human values connected with the newly won material wealth. The book consciously appraises the purposes and means of change to assure the highest human outcomes and serve a balanced social development.

Book Three

On The Nature of Cities: Toward Enduring and Creative Human Environments

While cities have always held leading roles in civilization, starting as mere encampments and gradually achieving excellence if not grandeur, modern American cities are initially build to a high standard but then seriously decline in a matter of decades. Built almost entirely to the giant scale of automotive capacities to access immense land areas and promote urban sprawl, cities today have reduced urban efficiency, promote escape from, and abandonment of, older areas. They have reduced livability in neighborhoods split by wide roadways and jammed by traffic. The results compromise or defeat much of the progress of the last century and claim much of the expanded industrial capacities. The book pleads for a new vision of urban form, one in which the city plays an active role in progress and becomes once again the active force and positive artifact of civilization in its own right.

Book Four

Forging a More Perfect Union: For a Grand Harmony of Cities, Democracy, Ecology

This book stresses how cities are the dynamic but largely ignored and seriously misconceived realm of modern society—a failure to evolve a humane urban theory and for them to be prized for their unique and immense potential. Consequently cities have evolved as simple containers of technology, economics, housing, and varied public activities, leaving the person massed yet fragmented, isolated, and alienated. They have been created as systems of waste and deterioration, causing many inefficiencies, escape, and abandonment. The book sets forth a theory of development, stressing cohesive communities, principles of their creation, and emphasizing their numerous combined physical, institutional, ecological, economic, social, and cultural benefits.

Book Five

The Runaway Economy: The Rhetoric Is Growth, The Issue Is Freedom

The book promotes a different way of thinking about economics, exposes the oppressive effects arising under the affluent market economy, and emphasizes how it grows upon the increasingly destructive outcomes of certain products, most notably the automobile. The concept also stresses the need for economics to be directed by a system of social goals established *outside* of economics while abandoning the increasingly meaningless economic growth arising from the closed thinking characteristic of economics.

Book Six
American Communities: The Next Human Advance, A New Class of Freedom

This essay proposes the conscious establishment of modern communities in urban areas with a clear physical and organizational form that is necessary to pursue human integrity and an expansive human fulfillment. A community thus unites and immensely expands the environmental, institutional, social, economic, and cultural possibilities close to the person. Being at the human scale, a major universal efficiency complements the social values of its large building and surrounding recreation areas, like those of a country club. Hence, the self-governing community expands the content of human sovereignty and develops a new scope of *positive social* democracy. As such, it promotes a distinct development of a new *public interest* sector of society, comparable to those of expanded democracy and redirected capitalism.

Book Seven
Forging a Supremacy of Person: For a New Human Liberation

The book's thesis is that political democracy and economic development are but early stages of a much greater human development, where the potential prize of society is the person itself. The concept of *virtu* is applied to the person as society's greatest end purpose and most dynamic means. Through programs that promote self-development, a new approach to learning, the creation of community, new realms of interpersonal and social dialogue, and economic transformation, a fundamental human liberation becomes possible.

ABOUT THE AUTHOR

Kenneth R. Schneider

Following service in the Marine Corps in World War II, Kenneth Schneider studied sociology and city planning at the University of California, Berkeley. He worked in planning for some years, including assignments at the United Nations in New York. Subsequently he worked with CARE in the Philippines, Sierra Leone, and Jordan. Following his overseas work, Schneider stressed his role as a generalist writing *The Destiny of Change* and *Autokind Vs. Mankind,* and completed *On the Nature of Cities* while founding a small business publishing large format post cards. He managed the company for almost twenty years. Since 1997 he has completed four additional books, noted in the front piece, all intended to stimulate a basic dialogue about fundamental change in society.

0-595-33817-8

www.ingramcontent.com/pod-product-compliance
Lightning Source LLC
Chambersburg PA
CBHW061344280526
45784CB00001B/121